SpringerBriefs in History of Science and Technology

The *SpringerBriefs in the History of Science and Technology* series addresses, in the broadest sense, the history of man's empirical and theoretical understanding of Nature and Technology, and the processes and people involved in acquiring this understanding. The series provides a forum for shorter works that escape the traditional book model. SpringerBriefs are typically between 50 and 125 pages in length (max. ca. 50.000 words); between the limit of a journal review article and a conventional book.

Authored by science and technology historians and scientists across physics, chemistry, biology, medicine, mathematics, astronomy, technology and related disciplines, the volumes will comprise:

1. Accounts of the development of scientific ideas at any pertinent stage in history: from the earliest observations of Babylonian Astronomers, through the abstract and practical advances of Classical Antiquity, the scientific revolution of the Age of Reason, to the fast-moving progress seen in modern R&D;
2. Biographies, full or partial, of key thinkers and science and technology pioneers;
3. Historical documents such as letters, manuscripts, or reports, together with annotation and analysis;
4. Works addressing social aspects of science and technology history (the role of institutes and societies, the interaction of science and politics, historical and political epistemology);
5. Works in the emerging field of computational history.

The series is aimed at a wide audience of academic scientists and historians, but many of the volumes will also appeal to general readers interested in the evolution of scientific ideas, in the relation between science and technology, and in the role technology shaped our world.

All proposals will be considered.

Leo Corry · Raya Leviathan

Chaim L. Pekeris and the Art of Applying Mathematics with WEIZAC, 1955–1963

 Springer

Leo Corry 🆔
Cohn Institute for the History
and Philosophy of Science and Ideas
Tel Aviv University
Tel Aviv, Israel

Raya Leviathan
Cohn Institute for the History
and Philosophy of Science and Ideas
Tel Aviv University
Tel Aviv, Israel

ISSN 2211-4564 ISSN 2211-4572 (electronic)
SpringerBriefs in History of Science and Technology
ISBN 978-3-031-27124-3 ISBN 978-3-031-27125-0 (eBook)
https://doi.org/10.1007/978-3-031-27125-0

This Springer imprint is published by the registered company Springer Nature Switzerland AG
The registered company address is: Gewerbestrasse 11, 6330 Cham, Switzerland

Acknowledgments

Work on this book has involved large amounts of archival research and personal interviews. In those cases where the archives provide keys to the documents, we have referred to the original archival key. We have also used documents that are undated or unsigned. In those cases, we simply cite by giving the available information. All the documents and photographs are cited and used by permission of the archives or holders in personal collections, as indicated in each case. We would like to thank and acknowledge those individuals and institutions, which have positively reacted to our enquiries, providing us with important information and insights, as well as permission to publish documents and photographs.

We thank the Weizmann Institute of Science Archives (WIA), Rehovot, Israel, for their enthusiastic cooperation. At the Department of Applied Mathematics (DAM) of WIS, we benefited from the kind help of its Head, Prof. David Peleg, as well as of the Administrative Assistants, Mr. Raanan Michael and Mrs. Carol Weintraub. They provided us with crucial access to the Chaim Pekeris Archive (CPA) and to the unpublished scripts of conversations of the department pioneers. The Weizmann Memorial Organization, "Yad Chaim Weizmann," at Rehovot, provided us with access to and important guidance at the Chaim Weizmann Archive.

We want to express our sincere and special thanks to Aviezri Fraenkel, Prominent WEIZAC Pioneer, for many informative and illuminating conversations. Likewise, we thank the late Gerald Estrin for sharing important information via email interchanges. Heartfelt thanks go to Ruth Riesel and Naomi Yaron, daughters of the late Zvi Riesel—another WEIZAC pioneer—for useful conversations and photographs taken from their personal collections.

We want to thank the scientists who shared with us their work experience with Pekeris: Zeev Luz, Leslie Leiserowitz, Hans Jarosch, Saul Abarbanel, Krzysztof Frankowski, Avihu Ginzburg, Yigal Accad, as well as Meir Weinstein and Gideon Eidelboim.

Estrin and Fraenkel provided access to their private copies of six DVDs with enlightening video interviews conducted with the WEIZAC pioneers at WIS in April 1983. Now, the videos are freely available on a webpage entitled "The Computer

Pioneers: Weizmann Institute Video Oral History". https://ethw.org/Archives:The_ Computer_Pioneers:_Weizmann_Institute_Video_Oral_History.

While working out the technical details of the scientific issues discussed in the book, we received insightful and friendly assistance from Inbal Oz, Yaron Oz, and Shmuel Marco, to whom we are indebted and to whom we sincerely thank. We also acknowledge the helpful comments by an anonymous referee on a previous version of the manuscript.

For their advice on preparing the final version of this book, we would like to thank Matteo Valleriani, Editor of *Springer Briefs in the History of Science and Technology*, as well as to Christopher Wilby and Arumugam Deivasigamani and their team at Springer.

We acknowledge the generous support of the Naomi Prawer Kadar Foundation, through the 2022 Kadar Family Award for Outstanding Research.

This research was supported by The Israel Science Foundation (grant No. 472/21).

We are happy to thank our families for their continued support and for the sympathy and appreciation that they always express toward our work.

Tel Aviv, Israel   Raya Leviathan
April 2023   Leo Corry

Contents

Abbreviations and Acronyms

ACM	Association of Computing Machines
AJP	Alterman, Jarosch, and Pekeris
CFL	Courant-Friedrichs-Lewy
CPA	The Chaim Pekeris Archive (DAM–WIS)
DAM	Department of Applied Mathematics, WIS
ETH	The Swiss Federal Institute of Technology, Zürich (Eidgenössische Technische Hochschule)
HMF	History of the Mathematics Faculty, WIS. Unpublished scripts of conversations of Lee Segel with the DAM pioneers: Chaim L. Pakeris (Feb. 24, 1987); Joe Gillis (Feb. 10, 1987); Zvi Riesel (Feb. 18, 1987). Kept at DAM secretariat
IAS	Institute for Advanced Study
ICC	International Computation Centre
IDF	Israel Defense Forces
INA	Institute for Numerical Analysis
INAC	Istituto Nazionale per le Applicazioni del Calcolo
ISA	Israel State Archive
MIT	Massachusetts Institute of Technology
NBS	National Bureau of Standards
NMR	Nuclear magnetic resonance
NORC	Naval Ordnance Research Calculator
NPL	National Physical Laboratory
ODE	Ordinary differential equation
PDE	Partial differential equation
RK	Runge–Kutta
SOVA	Smithsonian Online Virtual Archive, Computer Oral History Collection
UCLA	University of California, Los Angeles
WIA	Weizmann Institute Archive, Rehovot
WIS	Weizmann Institute of Science, Rehovot
WWI, WWII	World War I, II

Chapter 1
Introduction: WEIZAC, Pekeris, and Applied Mathematics

Abstract WEIZAC, the first electronic computer built at the Weizmann Institute in the recently established State of Israel, was operational between 1955 and 1963. The driving force behind the project was the applied mathematician Chaim Leib Pekeris, and the chief engineer of the project was Gerald Estrin, who had actively participated in the IAS computer project. For Pekeris the electronic computer was an instrument meant to serve the aims of a Zionist vision, in which applied science would play a fundamental role. WEIZAC would help to achieve technological progress in the young country and to enhance the skills of the local engineers. Pekeris paved the way to the insitutional approval of the project, mobilized financial resources, and put together a team of talented engineers and scientists, some of whom came to Israel especially for the task. The WEIZAC project, under his leadership, produced important scientific results and opened the way to the development of a very successful computer culture in Israel.

Keywords WEIZAC · Chaim Leib Pekeris · Gerald Estrin · Applied mathematics · Zionism

In the early 1950s, an unlikely project aimed at building an electronic computer in the recently established State of Israel was initiated at the Weizmann Institute of Science (WIS). The project was successfully completed in the years 1954–55, and the WEIZAC, as the computer was called, worked in full capacity for almost a decade. Its designers and builders used cutting-edge technology and achieved the highest benchmarks of computing performance at the time. Mathematicians and scientists at WIS and at other research institutions in Israel, as well as members of Israeli government organizations, used the computer to advance science in Israel and to spread the word of this new technology all over the country. The driving force behind the project was the applied mathematician Chaim Leib Pekeris (1908–1993). The computer was modeled after the famous machine of the Institute for Advanced Study (IAS) in Princeton, which had operated since 1952, and the chief engineer of the project was Gerald Estrin (1921–2012), who had actively participated in the IAS computer project.

L. Corry and R. Leviathan, *Chaim L. Pekeris and the Art of Applying Mathematics with WEIZAC, 1955–1963*, SpringerBriefs in History of Science and Technology, https://doi.org/10.1007/978-3-031-27125-0_1

In a recent publication, *WEIZAC: An Israeli Pioneering Adventure in Electronic Computing (1945–1963)*, we described in detail the process that led to the project, its planning and early stages, and its immediate contribution to creating a computer-savvy community of users within the scientific and industrial sectors in Israel (Corry and Leviathan 2019). While stressing the role of Pekeris as the visionary leader who helped the project come true, we located the discussion within the realm of the broader historical issue of the role of science and technology in processes of nation-building around the mid-twentieth century in general and in the case of the State of Israel in particular. We also provided a brief, but thoroughly documented, analysis of the impact of WEIZAC on actual scientific research in Israel and beyond. Our account revealed the astounding extent to which research based on calculations performed with WEIZAC (as well as with the two machines that followed it at WIS, GOLEM A-B) played an important role in the processes that turned WIS into the world-class, leading institution that it soon became. It also made clear the extent to which WEIZAC was at the heart of the creation of a relevant community of users of computing technologies in Israel—scientists, engineers, technicians, research institutions, and governmental branches.

The present book is intended as a complementary follow-up to our previous publication, and it is devoted to analyzing, in greater detail and within its relevant historical context, the specific scientific contributions of Pekeris and his collaborators that were based on calculations performed with WEIZAC. This account must necessarily address, in the first place, the issue of the rise and development of numerical analysis as an autonomous discipline of advanced mathematical research. Of particular interest are the events that began in the 1940s and eventually led to an overall transition to the electronic era. These topics are discussed in Chap. 2, which sets the stage for the three main following chapters, which discuss those works of Pekeris of greatest and longer-lasting scientific impact. They cover work on integral equations, particularly the Boltzmann equation (Chap. 3), on the helium atom and the calculations related to its ground-state energy (Chap. 4), and on geophysics and seismology, particularly around the issue of the free oscillations of the earth (Chap. 5). WEIZAC was also used for scientific computing in projects led by other scientists. Brief outlines of such projects are presented in Chap. 6. Finally, in Chap. 7, we present an overview of some of Pekeris's additional activities after the WEIZAC era and of some important developments at the Department of Applied Mathematics at WIS.

Each chapter opens with a brief historical account that provides the background and the historical context of the realms where Pekeris pursued his research, either alone or in collaboration. Each chapter also comprises an account of some of the main technical issues central to these realms. Such accounts, however, are not intended to provide a complete formal derivation of each equation or algorithm mentioned. We assume that the interested reader can follow all the details in the original publications, which we cite for all cases, while our account is meant to provide a roadmap for such reading. It is also possible to skip some of these details and yet continue to follow the overall coherence of the account for each topic. Still, we thought it convenient to allow any reader with some scientific background to get a clear idea of the kinds of considerations that led Pekeris and his collaborators to the point where numerical methods could be used and where WEIZAC would become the crucial tool for completing the tasks at issue. Such considerations comprise physical assumptions, mathematical derivations, attempts to simplify the algorithms, and others, which we explain in a schematic way in each relevant chapter.

In this introductory chapter, we provide a summary overview of the background to the WEIZAC project, of which the interested reader may find additional details in (Corry and Leviathan 2019). The WEIZAC project was developed in the sleepy farming town of Rehovot, about 20 km south of Tel Aviv, at an unlikely confluence of historical threads which involved—each in its own way—events with a lasting impact on a global scale over the second half of the twentieth century: the creation of the State of Israel and the rise of high-speed electronic computation.

The Weizmann Institute of Science was formally opened in 1949, but it had its origins in the Sieff Institute for Chemical Research in Rehovot, whose opening ceremony took place in April 1934. In his dual role of reputable scientist and prominent Zionist leader, Chaim Weizmann (1874–1952) was actively involved in the founding of two main academic institutions that were established in Mandatory Palestine: the Hebrew University in Jerusalem (officially opened in 1925) and the Weizmann Institute in Rehovot. The idea of establishing an institution of higher learning in Palestine, meant to serve as a spiritual center for the Jewish people, became a prominent issue in the deliberations of the Zionist congresses since they were first convened in Basel in 1897. Weizmann and Albert Einstein were among the enthusiastic supporters of this idea.

A pressing question that occupied the minds of those who were involved in the creation of these two institutions concerned the way in which they should and would contribute to the broader aims of the Zionist project, and particularly on the relative importance that should be accorded to theoretical (or pure, or basic) versus applied branches of science. Weizmann's views in this regard underwent interesting changes and were ambivalent throughout the years, but they came to play a crucial role in the process that led to the successful development of the WEIZAC project. Both Einstein and Weizmann explicitly opposed the idea of establishing a university with low academic level, meant to meet, above all, basic employment and subsistence needs of the Jewish population in Palestine. However, while remaining committed to the idea that theoretically oriented research conducted along the highest scientific standards would be crucial for the healthy development of a Jewish national entity in Palestine as envisioned in his Zionist views, Weizmann also acknowledged the immediate needs of the Jewish settlers and immigrants and the obligation to contribute to that direction as well. His eventual support for the WEIZAC project was strongly related to the former consideration—an electronic computer as a tool for cutting-edge scientific research—but also to the development of high-level technological capabilities in the newly created State of Israel.

WEIZAC, whose design was based on that of the IAS machine, was one of the earliest high-speed electronic, digital stored-program computers built in the world. The IAS machine implemented the principles of the so-called Von Neumann Architecture. By 1950, there were about ten automatic computing research centers around the globe, with less than one thousand people seriously involved in the use of this new technology, most of them in the US and Great Britain (Aspray 1985, ix). Within a few years, one dozen countries had computers, either working or under construction. In a survey on Automatic Digital Computers, based on information gathered in the beginning of 1953 and conducted by the Office of Naval Research, we find a list of nearly one hundred such machines, nationally distributed as follows: USA (~70), Britain (10), Germany (8), Japan (3), Canada (2), France (2), Holland (2), Sweden (2), Switzerland (2), Belgium (1), Australia (1), and Norway (1) (Blachman 1953). In 1955, there were already about 200 computers in about 15 countries (Aspray 1986).

Only developed countries could afford at that time the necessary resources and the appropriate technologically advanced industry needed for building and running such automatic computing machines (Cortada 2013). Obviously, the budding State of Israel did not belong to this league. Its economy was of a much smaller scale, and it continued to suffer from the traumatic impact of the recently finished war. In its early years, the Jewish population doubled itself in three years, and the Israeli society faced enormous challenges that involved defense as well as economic issues, together with the burden of absorbing a huge number of immigrants and refugees.

Years before the WEIZAC was built, Pekeris started to conceive in very concrete terms the research agendas and the kinds of problems that he wanted to address in various fields of applied mathematics as well as the kind of computing techniques he would apply in order to do so. Like many other scientists involved in similar research, Pekeris had been applying numerical methods of increasing difficulty and efficiency, and performing calculations that required ever more intense resources, long before electronic computers were available. Eventually, this background allowed a smooth transition to the electronic era.[1] He made sure that the electronic computer was built at WIS and, at the same time, he also became the chief consumer of computation time and resources, often working with collaborators who played key roles in his research projects, as we will see below. Among them were researchers who eventually achieved international prominence on their own, such as Zipora Alterman (1925–1974) and Philip (Pinchas) Rabinowitz (1926–2006), as well as some of Pekeris's PhD students, such as Hans Jarosch (1928–2013), Yigal Accad (b. 1936), Ivor Martin Longman (1923–1993), Hana Lifson (1920–2016), and Krzysztof (Kriss) Frankowski (1932–2021). All of them, and many others, spent many hours programming the WEIZAC for Pekeris, producing important results as we will see below.

Pekeris, a Lithuanian-born Jew, received his basic mathematical training in leading academic institutions in the USA. He became acquainted with electronic computers as early as 1943 and had the opportunity to develop his skills, hands-on, working with the most advanced machines that were operational at the time. He joined WIS in 1948, as head of the Department of Applied Mathematics (DAM), intent on pursuing

---

[1] Similar processes affected various scientific communities where computing-intense methods existed prior to the advent of electronic computers, and that made a successful transition to the new era. Two noteworthy cases are those of *pure* mathematics, mainly number theory (Corry 2010) and quantum chemistry (Park 2003).

a research agenda focused on the application of computing-intensive mathematical methods in a variety of scientific disciplines such as theoretical and nuclear physics, meteorology, oceanography, seismology, and others. His project for building an electronic computer was the flagship initiative at the center of this agenda, an agenda that was diametrically opposed to the ethos of pure mathematics that had been successfully established in Mandatory Palestine, beginning in 1925, at the Einstein Institute of Mathematics of the Hebrew University of Jerusalem (Corry and Schappacher 2010), and about which we say more below.

The academic background and unique scientific personality of Pekeris were perfectly adapted to the pursuit of advanced and broad-minded research in applied mathematics, with a strong component of numerical methods. His intimate acquaintance with current topics of interest and open questions in various physical branches was remarkable.[2] In 1925, Pekeris received a degree in mathematics from MIT, where he continued with graduate studies in the department of aeronautical engineering, specializing in meteorology, then a new discipline. He completed his master's degree in 1929 and his doctorate (Sc.D.) in 1933. His advisor was Gustav Rossby (1898–1957), the de-facto director of the American Meteorological Project and a world-leading meteorologist. Working in the department of geophysics at MIT until 1940, Pekeris established his position as a promising young scientist. He made creative contributions to geophysics, astrophysics, and hydrodynamics. These were all new fields of enquiry in applied mathematics, in which intensive computations played a central role, and to which Pekeris would continue to contribute over the years, to a large extent based on calculations conducted with WEIZAC and later on newer generations of computers.

During WWII, Pekeris was involved in military research at Columbia University, where he investigated the propagation of acoustic waves and pulses. The assignment at Columbia allowed him to work, hands-on, with a state-of-the-art electronic computer, the Relay Interpolator (later named Model II), that had been operational at Bell Labs in New York since September of 1943 (Irvine 2001). Pekeris in 1987 described retrospectively his first impression from the automatic computer with the following words:

> In the middle of the war there was a team of the Division of War Research at Columbia University located on 64th floor of the Empire State Building. One day I was told that we had available a device that does computations in Bell Telephone Research Laboratories, in downtown New York. I went down there and I saw a computer built on relays, mechanical relays ... The thing that impressed me most at first is how the most complicated mathematical operations can be built on these simple elements, yes or no. To this day it's a remarkable thing.[3]

After the war, Pekeris became head of the mathematical physics group at Columbia University. During these years, he was involved in additional research of seismic

---

[2] For a comprehensive list of Pekeris's publications, see https://www.weizmann.ac.il/math/sites/math/files/pekeris.pubs.pdf (accessed Oct. 7, 2021).

[3] Lee Segel, Conversation with Pekeris, Feb. 24, 1987 (HMF).

wave propagation. After the Office of Naval Research had established the meteorological project under the direction of John von Neumann (1903–1957) at the IAS in Princeton, Pekeris was invited in 1946 to participate as a consultant.[4] The initial aim of the project was to examine the potential use of the electronic computer in theoretical meteorology research and in weather forecasting (Harper 2008, pp. 98–121). Participation in this project deepened Pekeris's understanding of the digital computer's capabilities and strengthened his personal relationship with von Neumann.[5]

Von Neumann was the most prominent figure behind the IAS electronic computer project and one of the most influential leaders in science in the USA at the time. Beyond his many important works in central disciplines of pure mathematics since the beginning of his career in the early 1920s, during the war and immediately thereafter, he made seminal contributions in many computing-intensive fields of mathematical research. Pekeris and von Neumann corresponded frequently about the use of electronic computers for solving open problems in physics based on numerical methods. An interesting example is found in a letter of 1950, where von Neumann politely declined an invitation to visit Israel, allegedly for lack of time, while providing some details about difficulties found in the current work on the IAS machine, particularly in relation to the use of Williams cathode-ray tubes for the memory. Von Neumann also asked Pekeris about his own recent progress and the possible use of automatic computers in solving some outstanding problems:

> What you write me about your work in connection with the Schrödinger equation and the self-consistent field problems sounds very interesting. The differential equations which you give as determining the problem of Helium II do not look vicious. How bad are they from the numerical point of view? They should certainly not be bad for a high-speed computer. Quite apart from more ambitious devices, the ENIAC or the SSEC, or the Harvard Mark II or III should be sufficient here. What are your conclusions regarding the behavior of these equations from this point of view? It would certainly be good to discuss these and other similar problems personally when you are here.[6]

An often-cited anecdote, originally told by Estrin, referred to a remark by von Neumann when asked about plans to build a computer in the young State of Israel: "What will that tiny country do with an electronic computer?" Von Neumann's reported answer could not have been more precise, and in many senses prophetic:

---

[4] Pekeris to Ettlinger (University of Texas), Dec. 13, 1946 (CPA).

[5] Rubinoff said that "Pekeris in fact was a rather close friend of Johnny von Neumann's" (Rubinoff, interview by Richard R. Mertz, May 17, 1971. SOVA,1969–1973, 1977). A year before his death, von Neumann wrote to Pekeris (von Neumann to Pekeris, Feb. 24, 1956 (CPA)): "I am really touched by the signs of true friendship that you are giving me".

[6] Von Neumann to Pekeris, Feb. 23, 1950 (WIA 72–3/96). The SSEC (Selective Sequence Electronic Calculator) was an electromechanical computer built by IBM, operational between January 1948 and August 1952. It is remarkable that von Neumann mentioned it here in combination with the other two, since, although it possessed many of the characteristic features of a stored-program computer, it was not fully electronic. As a matter of fact, also the Mark II was electromechanical, relay-based and not stored- program (Haigh et al. 2016, pp. 247–252). Thus from the three computers mentioned, only ENIAC was fully electronic.

"Don't worry about that problem. If nobody else uses the computer, Pekeris will use it full time!" (Estrin 1991, p. 319).[7]

Pekeris's scientific horizons were impressively broad. He gave much thought to the question of the mutual relationships between pure mathematics, applied mathematics, and physics, and the unprecedented way in which they had drifted away from one another after WWI, even before the rise of the electronic computer. Already in 1945 he wrote:

> The experience of this war has given a great stimulus to Applied Mathematics in this country and in England. This is evidenced by (1) the establishment in 1940 of the Institute of Applied Mathematics at Brown University, (2) the publication of the new Journal of Applied Mathematics, (3) the opening of the Applied Mathematics Center at M.I.T., (of whose steering committee I am a member), (4) the recent establishment of Applied Mathematics Centers at Harvard and Columbia, (5) the formation of such centers at the leading industrial laboratories in this country, such as the General Electric Company, the Bell Telephone Laboratories, etc., (6) the installation in several Eastern Universities and in industrial laboratories as well as government research divisions, of computing machines, popularly known as "mathematical brain". (The latest such "brain" now under construction will cost more than $200,000.) 7) The tremendous expansion during the war of "differential analyzer" work under the direction of Professor Hartree of Manchester University.[8]

He considered this to have been a major revolution in science that had never received the attention it deserved. Ph.D. candidates in pure mathematics—he indicated with sorrow—could nowadays complete their studies without ever having to hear about the Second Law of Thermodynamics. Physicists, on the other hand, were lawgivers with an unexcelled record, but even in their orderly world one can never know exactly what is happening beneath the surface and, indeed, riots explode once in a while in their provinces. Pekeris thought of himself as belonging to a tradition whose emblematic representative was Lord Rayleigh (1842–1919), who had been at the forefront of physical research all his life and was also the leading authority in the field of applied mathematics. Like Lord Rayleigh, Pekeris thought that all natural phenomena are of equal interest, whether they occur in the laboratory, in the atom, in the atmosphere, in the sun, in the ocean or in space. Also like Lord Rayleigh, he declared in 1964 that his chosen audience was neither that of pure mathematicians nor that of physicists but rather "nature itself" (Pekeris 1964, p. 2). This was also the view promoted by von Neumann, about which we say more below.

When he joined the planning committee of WIS in 1946, Pekeris was guided by a clear vision of the "dire need of modernization both with regard to equipment as well as methods of manufacture"[9] in Palestine. The Weizmann Institute—and his own project for building an electronic computer within that institution—would play a leading role in the way to make this vision come true. Pekeris argued that

---

[7] In 1973, Estrin cited von Neumann as saying: "This guy Pekeris, is extremely versatile, hard-working guy.... I guarantee you that if nobody else in that country does anything to use that machine, he will keep it busy full-time by himself." (Estrin, interview by Robina Mapstone, Jun. 15, 1973. SOVA #96, Box 6, Folder 12).

[8] Pekeris to Leon Roth, Mar. 13, 1945 (WIA 3–96–98).

[9] Pekeris to Bergmann, Mar. 9, 1946 (WIA 3–22–46).

building an electronic computer would be an important project for the institute as well as "a typical example in which the commercial importance of skill and ingenuity is conspicuous."[10] For him the electronic computer was an instrument to achieve technological progress and to enhance the technical skills of the local engineers, as well as a tool to advance mathematics and science. He conveyed his ideas to Weizmann and he seems to have convinced him from quite early on.

Nevertheless, Pekeris's ideas were not unanimously welcomed and the decision to go ahead with the project at WIS met with many hurdles. Fulfilling his plans required, in the first place, gaining the active support of the key figures at the Institute, firstly Weizmann, and his closest associates, the chemist Ernest David Bergmann, (1903–1975), Scientific Manager of the Institute, and Meyer Wolfe Weisgal (1894–1977), Weizmann's personal assistant from 1940.

Pekeris intelligently built their faith in his ability to carry out the project successfully in the highly unfavorable conditions of the newly created state. In particular, through Weisgal's support, he was able to ensure the required funds. During the planning years of WIS, with Pekeris still in the USA, he directed his efforts at reaching the initial approval for the project. Later on, after his arrival at WIS, his actions were carefully planned to set the stage for the electronic computer in Israel. His strategy was successful by any standard. During the first years of the Institute (1949–1951), despite the adverse material conditions, Pekeris continued to push the project forward until the decision was finally reached and the computer became a reality.

Pekeris not only paved the way to a smooth approval of the project and created this sense of inevitability, he actually made the project viable by mobilizing financial resources, transferring knowledge, and recruiting the necessary professional personnel. He put together a team of talented engineers and scientists, some of whom came to Israel especially for the task. He found ways to acquire the necessary electronic equipment, some of which was hardly available in most countries of the world, and much less so in the entire Middle East. Particular difficulties arose in the attempt to purchase a Magnetic Core Memory to serve as the main storage unit in the machine. None of these tasks could be deemed easy to achieve, but Pekeris's incredible resourcefulness, combined with the complex network of connections that he had developed throughout the years, proved to be crucial and to allow for the successful realization of the project.

By the mid-1960s, Pekeris had already become a well-established and prominent scientist with a proven track record, and he thought that his applied mathematician colleagues would "not wholly reconcile themselves to this world until they can sit in their offices and, with the aid of electronic computers" solve certain open problems of which he gave some prominent examples. This included some to whose solution Pekeris himself had partially contributed with the help of WEIZAC. His list included the following (Pekeris 1964, p. 2):

---

[10] "The Following Examples", in the Pekeris-Bergmann correspondence file (WIA 24–74–6).

- Forecast the weather by numerical analysis,
- Predict the tides at every point of the world ocean on the basis of tidal theory alone,
- Predict the results of every aerodynamic measurement made on models in a wind tunnel, and with greater accuracy,
- Explain, on the basis of pure theory, every measured atomic and molecular spectral line,
- Determine the crystal structure of proteins from X-Ray photographs, and
- Be able to interpret the significance of every wrinkle in a seismograph in terms of the nature of the explosion (nuclear or earthquake) and the internal constitution of the earth.

"Clearly," he concluded in 1964 with a prophetic-sounding note, "there is no peace in the foreseeable future for applied mathematicians."

But even much earlier than that, while still at the hectic New York pace of Columbia and already planning his possible move to WIS, Pekeris had a rather detailed plan of what he intended to do upon arrival. Thus, for instance, in a letter of December 1946 to Herman J. Getzoff, a staff member of the American Committee of WIS, Pekeris outlined some of the projects that could be appealing for prospective sponsors of individual initiatives to be carried out at the institute. Some of these he described as being "important for the economy of Palestine," and in those cases he stressed "the utilitarian aspects of each project." He listed the electronic computer along with the design of other advanced scientific instruments for research in physics and chemistry that would find applications in industrial processes. High in the agenda that Pekeris suggested in 1946 as worthy of attracting potential donors in the USA were: (1) the development of geophysical prospecting for oil and water in Palestine and (2) a radiation laboratory for studying the properties of known and new substances in the microwave region. Pekeris emphasized the importance of conducting experimental studies about the feasibility of introducing microwave communications in Palestine and the possibility of entering the field of nuclear research.[11]

Alongside these more "utilitarian" kinds of projects, Pekeris also emphasized the importance of approaching "individuals ... in whom you could arouse more sympathetic attitude to a purely scientific project." To such persons, Pekeris advised that the following projects be mentioned: (1) seismological, gravimetric, and magnetic studies in the geophysical field; (2) the study of the ionosphere and of the structure of the upper atmosphere with the help of the planned electronic computer, and (3) the study of superconductivity in the radiation project. In the following decades, as we will see below in detail, Pekeris was able to achieve significant results in all of these projects and much more, while performing the necessary computations on WEIZAC.

By 1959, Pekeris and his collaborators were able to solve, with the help of WEIZAC, as many as 1078 simultaneous linear equations, and they maintained the lead in this regard for a few years. There were few machines operational at the time that could perform similar tasks. And to the extent that any existing machine came

---

[11] Pekeris to Getzoff, Dec. 8, 1946 (WIA 3–96/25).

close to such performance capabilities, there were surely few places in the world, if at all, where they could be thoroughly exploited for scientific purposes by a team of talented mathematicians working under a determined and knowledgeable leadership such as that of Pekeris at WIS.

In 1995, one year after Pekeris's passing away, the First Chaim Leib Pekeris Memorial Lecture was organized at WIS. The honor of delivering the lecture fell to Sir James Lighthill (1924–1998), Lucasian Professor of Mathematics at Cambridge University, a leading world expert in fluid mechanics and aerodynamics, and a long-time friend and collaborator of Pekeris. Lighthill described Pekeris as one of the twentieth century's pre-eminent figures in applied mathematics and credited him with the creation of one of the leading centers of "the art of applying mathematics" (Lighthill 1995, p. 11). This book is intended as an account of some of the main achievements of that center of excellence and of Pekeris's direct involvement, inasmuch as it concerns the period of time when WEIZAC was its main tool of computation.

References

Aspray, W. 1985. *Proceedings of a Symposium on Large-Scale Digital Calculating Machinery (1947). New introduction by William Aspray.* Cambridge, MA: The MIT Press.

Aspray, W. 1986. International Diffusion of Computer Technology, 1945–1955. *IEEE Annals of the History of Computing*, 351–360.

Blachman, M. N. 1953. *A survey of Auromatic Digital Computer 1953.* Washington D.C.: Office of Naval Research, Department of the Navy.

Corry, L. 2010. Hunting prime numbers–from human to electronic computers. *The Rutherford Journal—The New Zealand journal for the history and philosophy of science and technology* 3.

Corry, L., and R. Leviathan. 2019. *WEIZAC: An Israeli Pioneering Adventure in Electronic Computing (1945–1963).* Berlin: Springer.

Corry, L., and N. Schappacher. 2010. Zionist internationalism through number theory: Edmund Landau at the opening of the Hebrew University in 1925. *Science in Context* 23 (4): 427–471.

Cortada, J. W. 2013. How new technologies spread: Lessons from computing technologies. *Technology and Culture* 54 (2), 229–261.

Estrin, G. 1991. The WEIZAC Years (1954–1963). *IEEE Annals of the History of Computing* 13 (4): 317–339.

Haigh, T., M. Priestley, and C. Rope. 2016. *ENIAC in Action: Making and Remaking the Modern Computer.* Cambridge, MA: The MIT Press.

Harper, K. 2008. *Weather by The Numbers: The Genesis of Modern Meteorology.* Cambridge, MA: The MIT Press.

Irvine, M.M. 2001. Early digital computers at Bell Telephone Laboratories. *IEEE Annals of the History of Computing* 23 (3): 22–42.

Lighthill, M.J. 1995. *Ocean Tides from Newton to Pekeris. The First Chaim Leib Pekeris Memorial Lecture.* Jerusalem: Israel Academy of Sciences and Humanities.

Park, B.S. 2003. "The'hyperbola of quantum chemistry': The changing practice and identity of a scientific discipline in the early years of electronic digital computers, 1945–65. *Annals of Science* 60 (3): 219–247.

Pekeris, C.L. 1964. A brief history of the department of applied mathematics. *Rehovot, A Journal Published by the Weizmann Institute of Science and Yad Chaim Weizmann* 1–7.

Chapter 2
Numerical Analysis in the Age of Electronic Computing

Abstract The dawn of the digital era and the rise of the electronic computer signaled a most significant turning point in the history of the development and application of the discipline of Numerical Analysis. Methods for obtaining numerical results for differential equations related to specific domains of physics that had been introduced in the previous decades received now an enormous boost and were significantly improved and applied with increased success in an ever-growing spectrum of problems. Prominent among these were the Runge-Kutta method for solving ordinary differential equations (ODEs), and finite difference methods, such as the Courant-Friedrichs-Lewy (CFL), for partial differential equations (PDEs). They were crucial to the successfull aplication of WEIZAC to scientific problems. In addition, institutionalizing research and training in numerical analysis at the Weizmann Institute became a top priority for Pekeris. The person specifically commissioned with this important task was Philip (Pinchas) Rabinowinowitz.

Keywords WEIZAC · Chaim L. Pekeris · Numerical Analysis · Runge Kutta Method · Courant-Friedrichs-Lewy Method · Philip Rabinowitz

All of Pekeris's scientific achievements based on the use of WEIZAC comprise a sophisticated use of techniques developed over the first part of the twentieth century for obtaining numerical results for differential equations related to specific domains of physics. The dawn of the digital era and the rise of the electronic computer signaled a most significant turning point in the story of the development and application of such techniques and, indeed, for the discipline of Numerical Analysis. Such numerical methods as had been introduced in the previous decades received an enormous boost and were significantly improved and applied with increased success in an ever-growing spectrum of physical problems. The discipline and its associated institutions became emblematic of the processes that accompanied the transition from pre- to post-WWII science. In terms of individual mathematicians, the processes that marked the watershed whereby the discipline became inseparably identified with the advent of electronic computing are nowhere most clearly embodied—both at the material and the symbolic level—than in the iconic figure of von Neumann. This is true concerning both his seminal works and those he produced in collaborations with some leading contemporary figures. This is also true concerning the rise and consolidation of the

© The Author(s), under exclusive license to Springer Nature Switzerland AG 2023   11
L. Corry and R. Leviathan, *Chaim L. Pekeris and the Art of Applying Mathematics with WEIZAC, 1955–1963*, SpringerBriefs in History of Science and Technology,
https://doi.org/10.1007/978-3-031-27125-0_2

USA as the new world-leading center of science. And this is no less true concerning the displacement of the center of gravity in mathematics from the pure toward the more applied. It is perhaps even more significant, when it comes to the rise of a new kind of alliance between scientists and the main world centers of power, both economic and military (Dalmedico 1996, 2001).

Von Neumann's contributions in this regard were twofold. On the one hand, he was fully aware of the potential of open classical problems of mathematics as a source of challenging tests for the computing power of the new machines (as well as for the programming skills of those in charge of operating them). In 1949, for instance, he suggested using ENIAC to calculate values of π and e up to many decimal places (Reitwiesner 1950). Against the background of his interest at the time in questions related to randomness—and particularly in developing possible tests for checking randomness—he viewed the decimal expansions of these two numbers as useful sources of random sequences of integers where such tests could be initially tried.

On the other hand, von Neumann's war-related experience with practical problems requiring intensive computations led him to develop innovative mathematical tools especially adapted to making the most out of the new machines. Particularly influential were the Monte Carlo methods, initially introduced by Stanislaw Ulam (1909–1984) while working in the Manhattan project, and later perfected by von Neumann. These methods, which rely on the use of algorithms based on random samplings for obtaining numerical solutions of problems that are deterministic in principle, quickly found important applications in many fields, including physics, physical chemistry, and operations research (Richey 2010).

As Peter Galison has insightfully emphasized, the entire sets of skills that developed around the use of Monte Carlo methods belong to a "trading zone" between disciplines, where various practitioners—pure mathematicians, applied mathematicians, physicists, bomb builders, numerical analysts, industrial chemists, meteorologists, and fluid dynamicists—may have their own view about what they are all about. "Monte Carlo—Galison wrote—in some ways was the culmination of a profound shift in theoretical culture, from the empyrean European mathematicism that treasured the differential equation above all, to a pragmatic empiricized mathematics in which sampling had the final word" (Galison 2011, p. 151). But this important remark has to be strongly associated, above all, with the scientific personality of von Neumann. The classical idea of mathematizing nature—originating in the seventeenth century with the works of Newton, Laplace and their likes—underwent in the hands of von Neumann a meaningful transformation toward the idea of modeling, which became manifest, in particular, in the form of computer-based modeling (Israel and Millán Gasca 2009, pp. 57–76). In terms of domains of interest in mathematical physics, it was hydrodynamics that became for him paradigmatic in his quest for using computer-related methods of numerical analysis, particularly in addressing nonlinear problems. His wartime work on nuclear reactions led him to the realization that:

> ... many problems which do not prima facie appear to be hydrodynamical necessitate the solution of hydrodynamical questions or lead to calculations of the hydrodynamical type. It should be noted that it is only natural that this should be so since hydrodynamical problems

are the prototype for anything involving non-linear partial differential equations, particularly those of the hyperbolic or mixed type, hydrodynamics being a major physical guide in this important field, which is clearly too difficult at present from the purely mathematical point of view.[1]

Even before high-speed electronic computers had become a reality, by 1945 von Neumann was already convinced that "many branches of pure and applied mathematics are in a great need of computing instruments to break the present stalemate created by the failure of the purely analytical approach to non-linear problems" (von Neumann 1963, p. 2).

At the same time, once electronic computers started to become increasingly common, von Neumann expressed in a 1949 letter to Vannevar Bush (1890–1974), at MIT, his concerns about the fact that "many organizations are now asking for computing machines, although they will be completely unable to use them and have only the most amorphous ideas as to how those might be useful for them."[2] This negative judgement, to be sure, would by no means apply to Pekeris, whose attitude and expertise was more properly described by von Neumann's statement in the same letter, to the effect that:

> The entire computing machine is merely one component of a greater whole, namely, the unity formed by the computing machine, the mathematical problems that go with it, and the type of planning which is called for by both.

Of particular importance for our account here, concerning the methods that would become classical for numerical analysis in the electronic computer era, is one of von Neumann's remarkable papers, published in 1947 in collaboration with Hermann Goldstine (1913–2004). The paper and its sequel of 1951, that for some mark the beginning of modern numerical analysis, dealt with the all-important question of inverting matrices of higher orders, and the calculation of rigorous estimates of the errors that arise in such processes (Goldstine and von Neumann 1947, 1951). These articles are interesting not just because of the specific techniques they introduced, but also because of the way they discussed general issues of principle that are directly related to the use of electronic computers in this context, and in particular the question of the kind of errors inherent in it.

Von Neumann and Goldstine classified the possible types of primary errors (i.e., errors that may happen independently and that aggregate with one another as part of complex numerical calculations) into four categories. The first two types relate to the very idea of mathematically modelling a physical situation, an issue of fundamental importance for von Neumann's overall scientific outlook. Inherent to such models is the fact that only some phases of reality, and not reality itself, can be represented using them.[3] Hence, the following two types: (A) limitations derived from the kinds

---

[1] From a memorandum to Oswald Veblen. Cited in (Dalmedico 2001, p. 401).

[2] Von Neumann to Bush, Nov. 15, 1949. Cited in (Miklos 2005, p. 77).

[3] This concern is directly related to the idea of "Panmathematics" which, as explained in (Israel and Millán Gasca 2009, pp. 57–67), summarizes the gist of von Neumann's conceptions about mathematics and nature.

of idealizations, simplifications, and neglections that are unavoidable when mathematically formulating a physical situation; (B) observational errors in measuring the basic parameters of a given model.

In addition to these two, there is the problem (C) of approaching by elementary arithmetical operations that the computer can handle directly, and in a finite number of steps, transcendental functions like sin or log, as well operations like integration or differentiation. Any limiting process, "which in its strict mathematical form is infinite, must in a numerical computation be broken off at some finite stage, where the approximation to the limiting value is known to have reached a level that is considered to be satisfactory."

Von Neumann and Goldstine indicated, concerning errors of type (C), that their analysis applies to any kind of "computer" (Goldstine and von Neumann 1947, p. 1025):

> Digital computing by human operators, by 'hand' and by semi-automatic 'desk' machines, also computing by the large modern fully automatic, 'self-sequenced,' computing machines. Fundamentally, however, it applies equally to those 'analogy' machines which can perform certain operations directly, that are 'transcendental' or 'implicit' from the digital point of view.

Indeed, analog computers continued to be a viable, legitimate alternative to the idea of automatic computing well into the 1950s (Small 2013),and so was the idea that calculation-intensive tasks may be performed by groups of "human computers" (Grier 2013). Digital electronic computers as the dominating paradigm took some time to be established. It is thus only natural that von Neumann and Goldstine would refer to all of these alternatives, while pointing out that "the differences are only in degree (number of processes that rate as 'elementary' and 'explicit') but not in kind. Such differences, by the way, exist even among digital devices: one may treat square rooting as an 'elementary,' 'explicit' process, and another one not, and so on."

The fourth type of errors (D) mentioned in the article mentioned in the article complement those of type (C), as they refer to the fact that the machine cannot be fully rigorous and faultless at the level of the individual, "elementary operations" that they perform. Of course, the kind of "noise" and "rounding-off" problems that arise at that level in either digital or analog machines are somewhat different from each other, but still they arise in both cases.

These joint articles by von Neumann and Goldstine were highly influential and repeatedly cited, in conjunction with another collaboration of von Neumann (Bargmann et al. 1946), as well as with a no less famous article by Alan Turing (1912–1954) (Turing 1948). Given the many parallels between von Neumann and Turing in matters related to post-war design and implementation of electronic computers,[4] it is quite remarkable that Turing would also devote efforts to developing methods of matrix inversions. Turing's main concern at the time was, however, "with the theoretical limits of accuracy that may be obtained in the application of these methods, due to rounding-off errors." Working at the National Physical Laboratory in England,

---

[4] Such parallels, however, as argued in (Corry 2017), did not exist to the same extent in the pre-war period.

Turing had presented his plans for the ACE computer in 1946, the first model of which became operational in 1950. In November of 1947, he submitted for publication his method of matrix inversion, citing among others the paper of 1947 by Goldstine and von Neumann.

The war effort had brought about the accelerated development of automatic computing machines as well as of improved numerical methods for solving mathematical problems with their help. The various post-war articles co-authored by von Neumann and the one by Turing provide a clear indication of the growing awareness of the specific problems that would arise from the fruitful combination of these two aspects of the new disciplinary identity of numerical analysis. Pekeris's scientific education and early career were squarely embedded in this tradition, and his conception of the kind of mathematical research he intended for WIS derived directly from it. Before describing in some detail the place of numerical analysis within the activities developed at the Department of Applied Mathematics (DAM) of the Weizmann Institute in Rehovot, it is necessary to provide a brief account of two main aspects of the discipline as it developed at the time: (1) the main methods that became fundamental in all attempts at solving differential equations numerically and (2) the new institutions that were built to address issues related to scientific computing.

2.1 Methods

Two main families of numerical methods for solving differential equations were developed in the early decades of the twentieth century: (a) the Runge–Kutta (RK) method for solving *ordinary* differential equations (ODEs) and (b) finite difference methods, such as the Courant–Friedrichs–Lewy (CFL), for *partial* differential equations (PDEs). From the late 1940s on, both families of methods were increasingly improved and diversified so as to allow for adapted versions to be used with the new electronic digital computers.[5] RK-type methods figure prominently in the story of Pekeris's work with WEIZAC, and they deserve being briefly discussed here.

The first stage in the development of the RK-type of methods appeared in 1895 in a seminal paper by Carl Runge (1856–1927) (Runge 1895), right before he moved from Hanover to Göttingen. Runge was to develop a very distinguished and prolific career in Göttingen, becoming the first professor of "applied mathematics" in Germany. In the next few years, Runge's numerical methods were further developed, independently, by Karl Heun (1859–1929) (Heun 1900) and Wilhelm Kutta (1867–1944) (Kutta 1901).

---

[5] See the historical website of SIAM: "The History of Numerical Analysis and Scientific Computing." http://history.siam.org/. Nick Trefethen provides in his Oxford website a valuable, and fairly comprehensive, chronological list of names, dates, and important numerical algorithms, which are milestones in this long story: https://people.maths.ox.ac.uk/trefethen/inventorstalk.pdf. A more classical account appears in (Goldstine 2012). See also (Bultheel and Cools 2010; Butcher and Wanner 1996; Gear and Skeel 1990).

Fig. 2.1 Approximation by
Euler's method. The dotted
line is the unknown curve

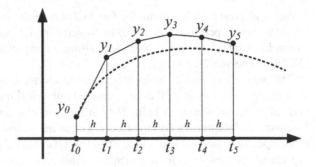

RK methods follow a "one-step" approach. They provide an approximate numerical solution to the given ODE, with approximations calculated stepwise, so that each iteration relies only on the latest calculated value. The basic step for calculating numerical values in Runge's method relies on the so-called Euler method for approximating values of a given function f, by means of a polygon which is built with the help of the derivative of f. Thus, given the initial condition indicated in Eq. (2.1):

$$y'(t) = f(t, y(t)), \quad y(t_0) = y_0, \tag{2.1}$$

we can approximate the values of f at a series of points in an interval, taking a fixed size h separating them: $t_n = t_0 + nh$. The successive approximations values, $y_n \approx y(t_n)$, are then given, as indicated in Fig. 2.1, by:

$$y_{n+1} = y_n + h \cdot f(t_n, y_n)$$

The RK-methods improve these approximations by calculating weighted averages over a large number of points. A straightforward example of this uses the weights $\frac{1}{6}, \frac{2}{6}, \frac{2}{6}, \frac{1}{6}$, as follows:

$$y_{n+1} = y_n + \frac{h}{6} \cdot (k_1 + 2k_2 + 2k_3 + k_4), t_{n+1} = t_n + h.$$

Here the coefficients k_i are given by:

$$k_1 = f(t_n, y_n),$$
$$k_2 = f\left(t_n + \frac{h}{2}, y_n + \frac{h}{2}k_1\right),$$
$$k_3 = f\left(t_n + \frac{h}{2}, y_n + \frac{h}{2}k_2\right),$$
$$k_4 = f(t_n + h, y_n + hk_3).$$

In this specific example, the approximation is said to be of order 4, in the sense that the order of the error e is $e \leq K h^4$, for some constant K. This way of taking weighted averages can be generalized in several obvious manners, around which questions of stability and accuracy have to be determined. A particularly important contribution to this thread of research came from Finnish mathematician Evert Johannes Nyström (1895–1960), who worked out the details of the fifth-order RK methods and, in addition, extended the methods to cover systems of second-order differential equations (Nyström 1925). Such systems are important in dealing with dynamical problems of the kind that arise in relation with the Boltzmann integral equation, about which we say more below.

An important development, specifically related with the question how to best implement RK-type methods in electronic computers, appeared in an article by Stanley J. Gill (1926–1975), working at the University of Cambridge (Gill 1951). Gill raised a rather interesting and original concern when he warned that processes which appear easily manageable for human computers may present particular challenges when trying to implement them in an automatic machine. In fact, this was one of Gill's many contributions to the overall development of programming techniques. He is credited, for example, with first introducing and implementing the general idea of a subroutine, in collaboration with Maurice Wilkes (1913–2010) and David Wheeler (1927–2004). They did so in their joint work with EDSAC, the first stored-program machine built in the UK, and operational between 1949 and 1958. Moreover, the three also co-authored what is arguably the first real textbook on computer programming(Wilkes et al. 1951).

Gill focused on existing methods for integrating differential equations by groups of human computers which rely on data accumulated through *several* stages in the process. Such methods are typically based on difference formulae; they have simple analytical forms and are easily remembered by a human computer. But Gill pointed at a specific difficulty that implementing the method in a machine would lead to: if a value y_{n+1} is found with the help of, say, y_n, y_{n-1} and y_{n-2}, then, in order to calculate, by a repetition of the same procedure, y_{n+2} from y_{n+1}, y_n and y_{n-1}, we must first replace y_{n-2} by y_{n-1}, y_{n-1} by y_n and y_n by y_{n+1}. Gill preferred an alternative approach that would be simpler from the point of view of writing the program and more rational in its use of the computer resources, particularly memory.[6] Thus we read in the introduction to his article:

> These shifting operations consume a serious proportion of the time and the instructions in a machine; the manual computer performs them merely by moving his eyes down the page. ... The number of storage registers available in any machine is limited, and if more than twenty or thirty simultaneous equations are being solved the shortage of registers may be serious. For n first-order equations, the number of registers required by any process is of the form $An + B$, which if n is large depends primarily on A. If several consecutive values, or backward differences of each variable are to be stored at any moment, A will be correspondingly large (Gill 1951, pp. 96–97).

---

[6] At the time, EDSAC had 512 18-bit words memory, a size which was later doubled.

The RK-type methods, accordingly, appeared to Gill and his collaborators as ideal for numerically integrating differential equations with electronic computers, given their character of one-step algorithms. He investigated various types of RK procedures involving Taylor expansions of the function f and up to the fifth power of the step h, but at the same time he also hinted at ways to expand his own analysis further on. He analyzed the conditions for ensuring that given a set of coefficients that define the weighted average in a four-stage RK method not only will be of order 4, but also that, even for large problems, the total memory requirement would not surpass four registers per equation. On the basis of a classification of fourth-order methods going back to Kutta, Gill explored most of the possible cases and found out that only for one of them could these aims be achieved. Gill also discussed how to implement the one-step computation so that not only the advantages of reduced memory requirements be attained, but also that the growth of round-off errors be kept as low as possible. In fact, he wrote the code for running the RK process on EDSAC, so as to allow for handling any number of equations within the storage capacity of the machine.

Also influential was an article by Robert Henry Merson (1921–1992), which followed a similar direction (Merson 1957). Merson, a scientist at the British Royal Aircraft Establishment, was motivated by the practical requirement of finding, along with a computed solution, a usable estimate of the truncation error committed within each step of the process. Merson extended Gill's ideas by focusing on a detailed analysis of the Taylor expansions for the exact and approximate solutions (McLachlan et al. 2017).The kind of approach developed by Gill and by Merson helped create an unwavering link between RK-type methods and electronic computers as the natural locus for numerical solutions of ODEs, and in particular as it came to be implemented in various projects that were assisted by WEIZAC.

Numerical approximations to solving partial differential equations (PDEs) were less common among the problems where WEIZAC was put to use. And yet some words are in order concerning the Courant–Friedrichs–Lewy method (CFL), which was the only significant one developed before WWII and later elaborated in the realm of electronic computing machines, particularly for solving PDE's of the hyperbolic type. The method was originally presented in 1928, in a paper written in Göttingen by Richard Courant (1888–1972) and his two young assistants Karl Friedrichs (1901–1982) and Hans Lewy (1904–1988) (Courant et al. 1928). Courant left Germany in 1933, and, after one year in Cambridge, UK, he emigrated to the USA in 1936, establishing himself at New York University (NYU), where he created the foremost center for applied mathematics that later on was called after him. Friedrichs and Lewy followed him to the USA, with Lewy developing his distinguished career at Berkeley.

In their 1928 article, they used the method of finite differences in order to demonstrate the existence of solutions of PDEs. In doing so, they discovered the phenomenon of numerical instability, whose enormous importance for the development of new methods became increasingly clear with time and was also investigated for RK methods. Focusing on equations related to the flow of a compressible, non-viscous fluid, they constructed numerical solutions by discretizing the space and time variables and solving the ensuing finite equations. The question of how to properly discretize is a subtle one in itself, but CFL soon realized that when we replace a certain type of PDE with a set of numerical approximations, arbitrarily close to the given equation, the resulting solutions may still be quite distant from the actual solution of the initial problem. Indeed, they showed that in order for the difference equations to converge to the original one, the difference scheme must use all the information contained in the initial data that effectively influences the final solution. This general requirement may be translated into a more precise condition, which came to be known as the CFL condition, namely, that the ratio of the spatial discretization step to the time discretization step must be smaller than a number C, which depends only on the specific differential equation and its initial conditions. Failure to meet this requirement will quickly lead to serious instability phenomena in calculating the numerical solutions. Obviously, this idea became of the utmost importance when algorithms for solving PDEs were implemented in electronic computers. Also in this realm, von Neumann played a crucial role when, as part of his work at Los Alamos, he reformulated the criterion in heuristic terms that could be put to work in a much more practical manner (Charney et al. 1950; Dalmedico 1996).

2.2 Institutions

The development of the discipline of numerical analysis and the eventual application of its methods by Pekeris and his collaborators cannot be understood without describing the role of a series of institutions devoted to that kind of pursuit. We mention, in the first place, the National Applied Mathematics Laboratory (later known as the Applied Mathematics Division—AMD), which was established on July 1, 1947, as a separate division of the National Bureau of Standards (NBS) of the USA Commerce Department. John H. Curtiss (1909–1977) was appointed as its head. A Harvard PhD in mathematics, Curtiss joined the NBS in 1946 as an assistant to the director, the distinguished nuclear physicist Edward Uhler Condon (1902–1974). Initially, Curtiss became responsible for all statistical issues handled at the NBS. In 1947, he also became the first president of the Association of Computing Machines (ACM). Working at the NBS during the years 1946–1953, Curtiss came to play a vital role in the development, procurement, and widespread application of computers in the USA (Lee 1995). Among other things, he realized the importance of turning the then budding mathematical field of numerical analysis into a central field of interest for professional mathematicians in the USA and beyond.

As initially conceived by Curtiss in 1947, the AMD was to have four sections:

1 Institute for Numerical Analysis (INA),
2 Computation Laboratory (CL),
3 Statistical Engineering Laboratory (SEL),
4 Machine Development Laboratory (MDL).

Curtiss undertook a massive recruitment program to implement the aims of the AMD. Given his extensive network of academic contacts and his own impressive list of achievements, he was certainly the right person to do the job. Besides, the timing could not have been better, given the recent developments in automatic electronic computing and the fact that many mathematicians were being demobilized from their WWII activities—activities where they accumulated unprecedented amounts of experience in applied mathematics.

The first mathematical appointees to INA were John Todd (1911–2007) and Olga Taussky-Todd (1905–1996). The first two projects undertaken at INA were: (a) "Characteristic roots of matrices," and (b) "Applications of automatic digital computing machines in algebra and number theory." The main aim of the institute was to help large numbers of professional mathematicians become proficient with the use of computers in solving a wide variety of difficult problems. As it was well funded, the INA was able to attract first-class mathematicians on sabbatical leave from their home institutions. In doing so, it definitely contributed to establishing a strong tradition of numerical analysis in the country. Another important task was to provide expert computing services to government agencies and research groups, which were involved in tasks that required solving difficult problems in applied mathematics (INA 1951).

The INA also continued with the long-established tradition of table-making that characterized the NBS for many years. During the economic depression of the 1930s, the Works Progress Administration set up a Mathematical Table Project in New York City under the scientific direction of the NBS (Grier 2003). Initially, calculations in INA were carried out on standard desk calculators. Only gradually punched card methods and, more generally, automatic computing techniques were widely adopted. The NBS was not allowed to compete with industry in building its own electronic computer, but noncompetitive experimental models were allowed. Two important, early electronic computers came out of these efforts. The first was Standards Western Automatic Computer (SWAC), built under the direction of Harry D. Huskey (1916–2017). Its construction began in January of 1949, and it became operational in August 1950, working until 1967. The second one was Standards Eastern Automatic Computer (SEAC). Modeled after the EDSAC and built under the direction of Samuel Nathan Alexander (1910–1967), SEAC was dedicated on June 20, 1950 (Kirsch 1998).

On April 7, 1950, Todd ran his first program on SEAC. It was meant to solve the Diophantine equation $ax + by = 1$, where a and b were originally taken to be the largest pair of consecutive Fibonacci numbers that fitted into the machine (< 244). One day earlier, Franz Alt (1910–2011), then Deputy Chief of AMD, had run a factorization program using a small sieve (Todd 1990).

But the INA was not the only institution related to numerical analysis that was established at this time. In the USA, the Rand Corporation—created in 1948, and partly funded by private sources, corporations and academic institutions—was devoted to the study of game theory, linear programming, and operations research, particularly with a view to applications in the military and related fields. The term

"Numerical Analysis" was used in the Whirlwind project as early as April 1947.[7]
A course named "Machine Computation" appeared in the MIT catalogue of 1949–
1950, and it included a "systematic study of topics in numerical analysis." In the UK,
in 1945 the National Mathematics Laboratory (NML) was created in parallel to the
NPL (National Physical Laboratory). Its first director was John Ronald Womersley
(1907–1958). All of these institutions worked in close collaboration with each other
(Hestenes and Todd 1991).

Also, in post-war USSR, the scientific establishment realized the need to develop
a new kind of mathematical knowledge based on the use of computer-oriented techniques. In a "Five Years Plan for Science" presented in 1947 by the President of the
Soviet Academy of Sciences, Sergey Ivanovich Vavilov (1891–1951), he addressed
this issue with the following words:

> Calculating machines have been known for centuries. But never has "machine mathematics"
> reached such scope as at the present time. New calculators devised on electrical princi-
> ples make it possible to solve extremely complex mathematical problems connected with
> technique and the various branches of natural science. So important do we think this side
> of mathematics is, that we are proposing in the immediate future to devote to it a special
> institute.[8]

The first academic meeting specifically devoted to topics in numerical analysis
took place at MIT on October 29–31, 1945. It brought together 84 participants from
the UK and the USA. It was organized by the National Research Council Committee
on Mathematical Tables and Other Aids to Computation, and it was chaired by Leslie
John Comrie (1893–1950), a prominent pioneer of scientific computation. Comrie
had headed several important scientific projects, such as the Computing Section
of the British Astronomical Association, between 1920 and 1922 (Croarken 2003).
In the opening address, it was stated that given the recent developments in the war,
automatic calculating machines were now available that "promise astronomical speed
for numerical processes" (Archibald 1946).

But probably the most visible milestone that announced the establishment of the
discipline was the "Symposia on Modern Calculating Machinery and Numerical
Methods," held on July 29–31, 1948, at UCLA. More than 500 registered atten-
dees participated in this event, sponsored by the INA and the NBS, as well as by a
series of additional academic and governmental agencies such as the departments
of astronomy, engineering, and mathematics at UCLA, the American Mathematical Society, the Association of Computing Machinery, the American Society of
Mechanical Engineers, the Institute of Radio Engineers, and many others (INA 1951).

The introductory article to the proceedings of that conference helps gain an idea
of contemporary conceptions about numerical methods in mathematics and some
important open questions at the time. The text was written by Douglas Hartree
(1897–1958), a British mathematical physicist with a unique background in calculational methods. During WWI, Hartree was involved in calculations related to ballistic

---

[7] See, for example, Forrester, Jay W., "Digital Computers in Science", Apr. 9, 1947 (https://dome.
mit.edu/ http://hdl.handle.net/1721.3/38896).

[8] Quoted in (Hestenes and Todd 1991, p. 3).

trajectories. Later on, as a student at Cambridge, he developed numerical methods for solving differential equations related to quantum mechanics, about which we say more below. He did much of this work with pencil and paper and later on with analog machines. Following a visit to MIT in 1933, where he became acquainted with Bush's differential analyzer, he constructed his own machine in Manchester using Meccano pieces. Then, during WWII and in its aftermath, Hartree got hands-on experience with the newest automatic calculating machines, and in 1946, he was appointed professor of mathematical physics at Cambridge. His inaugural lecture was entitled: "Calculating Machines: Recent and Prospective Developments and their impact on Mathematical Physics."

Hartree was at the time one of the most authoritative persons for assessing the state of the art in numerical methods (Darwin 1958). The list of unsolved problems in numerical analysis that he mentioned in his lecture included the following: elimination of approximately known roots of polynomial equations; solutions of systems of simultaneous nonlinear algebraic equations; relaxations methods for handling differential equations, both ordinary and partial, with given boundary conditions; eigenvalue problems in ordinary differential equations. In all cases, he emphasized the physical problems associated with the calculations for which numerical methods should be further developed. Thus, for instance, he spoke of solving "algebraic equations," meaning anything opposed to "differential equations," and his example was:

$$e^x = x + y$$

This equation relates to a significant physical situation, namely the analysis of crystal structures by means of X-rays (Hartree 1951, p. 2). Although in many, if not all, cases the solution of a set of equations can be handled without knowing the physical situation to which they refer, he stressed the physical context is helpful in suggesting means for their treatment:

> This is true, for instance, in the situation where a distribution of electrical charge, which can be treated as continuous, is moving under the influence of the mutual forces between its parts so that the field acting on any one charge depends on the distribution on all other charges, which themselves are not moving. This involves two kinds of equations: the equation of motion of the individual charges and Poisson's equation for the field arising from the space-charge distribution. (p. 8)

Hartree's point in explaining this was that certain "curious" boundary conditions may arise for which determining the solutions of the Poisson equation may require combining data about the emission current distribution, and this may affect the way in which the algorithms for solving the problem numerically need to be designed.

Yet another institutional context to be mentioned here is the International Computation Centre (ICC), formally established in Rome in 1951, under the sponsorship of UNESCO, as part of the post-war recovery plans for science in Europe. When plans started to be discussed for the center, some prominent institutions expressed their interest in participating. This included the Mathematisch Centrum in Amsterdam, and the Institut für Angewandte Mathematik at the ETH, Zürich. But the choice went to the Istituto Nazionale per le Applicazioni del Calcolo (INAC) in Rome, under the leadership of Mauro Picone (1885–1977). As initially conceived, the ICC was expected to provide a model for a worldwide network of international research laboratories to be created by the United Nations in the years to come. Conflicting interests of the parties involved, however, including the USA, translated into a delay of more than ten years before its convention was ratified. Also, Israel had been invited to participate in this initiative, and the leadership of WIS thought that this would be a useful venue to help address some of the institute's computational concerns, but Pekeris rejected this possibility out of hand.[9]

When the center opened officially, in January 1962, most Western European countries had already established their own institutions of scientific computation, as well as local computer industries, and hence the center failed to attract the kind of attention that its promoters had initially envisioned for it (Nofre 2014). Starting more than twenty years prior to the widespread introduction of electronic computers, and particularly productive under Mussolini's regime and in the war effort, Picone and his collaborators at the INAC, such as Carlo Miranda (1912–1982) and Sandro Faedo (1913–2001), had developed original numerical methods with applications in various fields of mathematics. Picone's own methods for the solution of PDEs may have had relatively little direct influence on modern numerical analysis, but his institutional and personal influence was definitely crucial to the field (Benzi and Toscano 2014).

Still under the denomination of "provisional," the ICC organized a symposium in Rome in September 1960 that became another milestone in the history of the discipline, devoted to the "numerical treatment of ordinary differential equations, integral and integro-differential equations." Pekeris, together with Alterman and L. Finkelstein, presented there their current work on the Boltzmann equation (Pekeris et al. 1960), about which more is said below.

2.3 Human Challenges

While a good acquaintance with the relevant physical contexts was for Hartree an important asset for anyone intent on contributing to the new field of numerical analysis, he also mentioned three other, quite different challenges that in his opinion affected the discipline at the time, coming from quite different sources. First is the following:

> One of the unsolved problems of numerical analysis is how to overcome the attitude of the mathematical fraternity toward the subject—an attitude exemplified by the comment of a distinguished mathematician introducing a lecture of mine on the mechanical integration of differential equations, that he had always regarded the solution of differential equations "a very sordid subject." (Hartree 1951, pp. 14–15)

---

[9] Pekeris to Frei, Jan. 8, 1952 (CPA).

The second one concerns the issue:

... of getting what I have called a "machine's eye-view" of a problem as presented to an automatic machine. It must be remembered that the machine will carry out the instructions given to it literally and blindly with no exercise of intelligence beyond these instructions. In doing a numerical calculation by hand process, one uses one's own intelligence, almost unconsciously, very much more than one realizes. ... Therefore, in programming a problem, all the unusual situations that may arise in the course of the solution of the problem must be anticipated, and the machine must be given adequate instructions to identify each one and to take the appropriate action if any one or any combination of them occurs.

The third one, also at the psychological level, is the problem:

... of getting enough "feel" for how the calculation is going when it is being done by an automatic machine. If one is actually handling the numbers oneself, one has a feeling for how the work is going which is difficult to get from seeing the completed results of the work of someone else and which seems almost impossible to get if the mechanism does the details of the work and never even exhibits them. My own experience with the differential analyzer has been that even a solution on this machine is too automatic to permit one to get a real feel for the way the calculation is going; on a problem of a new kind it has almost always been worthwhile to carry out the evaluations of one solution by hand myself to get a feel for the relative magnitudes of the variables, and the way in which they behave, before turning it over to the machine.

This interesting testimony probably reflects real challenges that in the early 1950s, the mathematical community at large, as well as the individual persons involved in this discipline, had to acknowledge and eventually overcome. As we will see shortly, Pekeris was not the kind of person that needed to be preached to in this regard. His specific abilities and the perspective from within which he intended to pursue applied mathematics at WIS, put him in a position where challenges such as mentioned by Hartree would not be an issue at all. Still, it is interesting to notice that all people involved in calculating with WEIZAC did pay focused attention to the issue of program reliability. In her regular progress reports of the years 1956–1957, Alterman provided detailed information to Pekeris on how the various programs that were being developed were constantly monitored in all their aspects, both the main program and the subroutines. Quite often, all calculations that WEIZAC was expected to perform were also reproduced separately with the help of desktop, mechanical machines.[10]

Richard Courant was likewise aware of the kinds of the challenges facing the discipline when it came to the human factor. He raised this point specifically in his keynote address to the ICC Rome meeting of 1960 (Courant 1960). Ever since his arrival in the USA in 1936, Courant had invested great efforts in building a center of research in applied mathematics at New York University. He was certainly a natural candidate for wholeheartedly adopting and then promoting the use of electronic computers at the core of mathematical research, and the interesting point is that by 1960 he still found it necessary to explain the reticence of important parts of the

---

[10] Alterman, "Progress Reports for the Years 1956–1957" (CPA).

mathematical community to do so. Among the interesting points he raised in his talk, the following are worth quoting here:

> Far too few young people of high caliber with enterprising minds and broad knowledge are prepared for the era of computing machines. … So far, one must admit that the great impetus which pioneers such as John von Neumann and Enrico Fermi imparted to the field has not caught up and stimulated enough of the younger scientific generation. … Why are so relatively few highly talented young scientists attracted by the fascination of computers?
>
> One of the reasons is, of course, that scientists who flourish as individual workers without time clock and rigid schedule are reluctant to give up such freedom and adjust their schedules to the precisely defined hours or even minutes demanded by the operation of a big machine. In astronomy or in high energy physics and in many activities in which the work depends on the cooperation of a team, such discipline seems to be natural and widely accepted. In the mathematical sciences it is an innovation which causes instinctive psychological resistance ….
>
> I think that numerical analysis should be recognized as a vital branch of mathematics. Many of the great mathematicians have taken a strong interest in this field. Gauss was perhaps foremost among them. … Now the advent of high-speed computers has opened vast new possibilities. Of course, such a development is still in its early stages. Its scope is much wider, its challenge much more exciting than many people seem to realize. At present, subjects such as matrix inversion and finite difference methods are in the focus of attention, and the interest is largely concentrated on problems arising from mechanics of continuous media. But indeed these domains have been cultivated primarily only by chance or immediate necessity ….
>
> The great progress in science after the French Revolution was largely the policy of combining teaching with research. I would say now that further progress can be made in the immediate future by combining teaching and research with numerical experimentation.

This quotation by Courant and, more generally, all the developments mentioned in this and the previous section should give us a clearer idea of the originality and boldness of Pekeris's undertaking at WIS. He was certainly not the kind of mathematician that would be "reluctant to give up" some kind of perceived freedom in the way he thought mathematics had to be practiced. In fact, one of the leading mathematical and administrative figures at WIS, Joseph Gillis (about whom we say more below) expressed, if anything, his concern for what he saw as possibly excessive enthusiasm with the electronic computer at DAM. He worked actively to "… keep in front of the younger people the fact that computing is neither the whole of mathematics or [sic] a substitute for it."[11] It is therefore so remarkable to see how focused and successful Pekeris was in leading his department under a well-conceived, but far from self-evident, scientific ideology that required enormous human and financial resources and the full confidence of the leadership of WIS if it was to materialize into significant scientific achievements. In the final account, no doubt, what Pekeris and his collaborators achieved fully justified the investment and the trust bestowed upon him.

2.4 Numerical Analysis with WEIZAC

Institutionalizing research and training in numerical analysis at WIS was certainly a top priority for Pekeris when he took the position of head of DAM in 1948. The person

---

[11] Pekeris to Gillis, Oct. 21, 1956 (WIA 6-74-24).

specifically commissioned with this important task was Philip (Pinchas) Rabinowitz. Rabinowitz was born in Philadelphia and got his PhD degree in Mathematics in 1951 from the University of Pennsylvania. In the years 1948–1949, Rabinowitz joined the mathematics group of the Whirlwind Computer Project at the Servomechanism Laboratory at MIT, which became operational in 1951 and is widely considered to have been the first digital electronic computer that implemented the idea of real-time output (Redmond and Smith 1980). This was his first experience in a cutting-edge project of digital computing (Davis and Fraenkel 2007). After completing his studies, Rabinowitz joined the Computation Laboratory of the NBS at Washington, DC, where he became an early user of the SEAC (Rabinowitz 1952). With this impressive record in the early stages of his career, and following an invitation to join the WEIZAC team, in mid-1955 Rabinowitz moved to Israel and became its chief programmer. He wrote the first utility programs and built up the scientific software library, in the form of subroutines, which would become the fundamental infrastructure for numerical solutions of most of the mathematical problems solved with the help of WEIZAC (Davis and Fraenkel 2007).

"Numerical Analysis" was formally mentioned for the first time as a separate discipline in WIS in the scientific report of 1956–1957, which included the following report:

> The high productivity of the electronic computer made it necessary to effect improvements in existing methods of numerical analysis, as well as to invent new ones.[12]

Rabinowitz taught the first course in programming and numerical analysis at WIS in 1956. It was a special course intended for future users of WEIZAC, either from WIS or from other institutions.[13] In November 1958, when the Weizmann graduate school in natural sciences started to operate, *Numerical analysis and programming* was one of the special courses taught in both semesters.[14] Pekeris was troubled by the fact that Rabinowitz was too occupied in support tasks on behalf of other WIS scientists who were using WEIZAC, and thus had no time for his own research.[15] However, Rabinowitz actually continued to do research well into the 1990s, alongside his teaching of numerical analysis. Two books he co-authored on the subject, *First Course in Numerical Analysis*, with Anthony Ralston (Ralston and Rabinowitz 1978), and *Methods on Numerical Integration*, with Philip Davis (David and Rabinowitz 1984), became standard and went through several editions. His contribution to the field of numerical integration was seminal (Gautschi 1995). One of his students at WIS, Nira Dyn, became a professor at Tel Aviv University. There can be no doubt that he became a central figure in the field (Bultheel and Cools 2010),[16] both at the

---

[12] Scientific Activity Report, 1956–1957, Sep. 1958 (WIA).

[13] Contribution of the Weizmann Institute to the Advanced of the Science of Electronic Computing 1961, Jun. 5, 1961. The document is not signed. Probably, Pekeris wrote it (CPA).

[14] Scientific Activity Report, 1958–1959 (WIA).

[15] Minutes of the meeting of the Scientific Committee held on Oct. 28, 1957 (WIA 8-9-6/76).

[16] In the preface to their *Birth of Numerical Analysis*, they wrote: "Another name we would have placed on our list as a speaker for his contributions to numerical integration would have been Philip Rabinowitz (1926–2006) had he still be among us."

level of developing new methods and assisting researchers in all fields at WIS in their use of WEIZAC. Thus, Rabinowitz was a crucial figure in helping Pekeris fulfill his vision for the DAM.

References

Archibald, R.C. 1946. Conference on advanced computation techniques. *Mathematical Tables and Other Aids to Computation* 2 (14): 65–68.

Bargmann, V., D. Montgomery, and J. von Neumann. 1946. Solution of linear systems of high order. Report prepared under contract NORD 9596 with the Bureau of Ordnance, Navy Department. Institute for Advanced Study.

Benzi, M., and E. Toscano. 2014. Mauro Picone, Sandro Faedo, and the numerical solution of partial differential equations in Italy (1928–1953). *Numerical Algorithms* 66 (1): 105–145.

Bultheel, A., and R. Cools. 2010. *The Birth of Numerical Analysis*. Toh Tuck Link, Singapore: World Scientific Publishing Co., Pte. Ltd.

Butcher, J.C., and G. Wanner. 1996. Runge-Kutta methods: Some historical notes. *Applied Numerical Mathematics* 22 (1–3): 113–151.

Charney, J.G., R. Fjørtoft, and J. von Neumann. 1950. Numerical integration of the barotropic vorticity equation. *Tellus* 2 (4): 237–254.

Corry, L. 2017. Turing's pre-war analog computers: The fatherhood of the modern computer revisited. *Communications of the ACM* 60 (8): 50–58.

Courant, R., K. Friedrichs, and H. Lewy. 1928. Über die partiellen Differenzengleichungen der mathematischen Physik. *Mathematische Annalen* 100: 32–74.

Courant, R. 1960. General problems confronting computer centers. In *Symposium on the Numerical Treatment of Ordinary Differential Equations, Integral and Integro-Differential Equations*. Basel: Birkhäuser.

Croarken, M. 2003. Table making by committee: British table makers 1371–1965. In *The History of Mathematical Tables: From Sumer to Spreadsheets*, ed. R. Campbell-Kelly, M. Croarken, and E. Robson, 235–267. Oxford: Oxford University Press.

Dalmedico, A.D. 1996. L'essor des mathématiques appliquées aux États-Unis: l'impact de la seconde guerre mondiale. *Revue d'histoire des mathématiques* 2(2): 149–213.

Dalmedico, A.D. 2001. History and epistemology of models: Meteorology (1946–1963) as a case study. *Archive for History of Exact Sciences* 55, 395–422.

Darwin, G.H. 1958. Douglas Rayner Hartree, 1897–1958. *Biographical Memoirs of Fellows of the Royal Society* 4: 102–116.

David, P.J., and P. Rabinowitz. 1984. *Methods of Numerical Integration*. Orlando, FL: Academic Press.

Davis, P.J., and A.S. Fraenkel. 2007. Remembering Philip Rabinowitz. *Notices of AMS* 54(11): 1502–1506.

Galison, P. 2011. Computer simulations and the trading zone. In *From Science to Computational Science*, ed, G. Gramelsberger, pp. 118–157). Zürich: Diaphanes.

Gautschi, W. 1995. The work of Philip Rabinowitz on numerical integration. *Numerical Algorithms* 9 (2): 199–222.

Gear, C., and R. Skeel. 1990. The development of ODE methods: A symbiosis between hardware and numerical analysis. In *A History of Scientific Computing*, ed. S. Nash and S.G. Nash, 88–105. Reading, MA: Addison-Wesley.

Gill, S.J. 1951. A process for the step-by-step integration of differential equations in an automatic digital computing machine. *Proceedings of the Cambridge Philosophical Society* 47: 95–108.

Goldstine, H.H. 2012. *A History of Numerical Analysis from the 16th Through the 19th Century*, vol. 2. New York: Springer Science & Business Media.

Goldstine, H.H., and J. von Neumann. 1947. Numerical inverting of matrices of high order. *Bulletin of the American Mathematical Society* 53 (11): 1021–1099.

Goldstine, H.H., and J. von Neumann. 1951. Numerical inverting of matrices of high order. II. *Proceedings of the American Mathematical Society* 2(2): 188–202.

Grier, D.A. 2003. Table making for the relief of labour. In *The History of Mathematical Tables: From Sumer to Spreadsheets*, ed. M. Campbell-Kelly, M. Croarken, R. Flood, and E. Robson, 265–292. Oxford: Oxford University Press.

Grier, D.A. 2013. *When Computers Were Human*. Princeton: Princeton University Press.

Hartree, D.R. 1951. Introduction. In I. f. (U.S.), *Problems for the Numerical Analysis of the Future*. Papers presented at the Symposia on Modern Calculating Machinery and Numerical Methods, pp. 1–12. US Government Printing Office.

Hestenes, M.R., and J. Todd. 1991. *NBS-NIA, the Institute for Numerical Analysis, UCLA 1947–1954, NIST Special Publication 730, National Institute of Standards and Technology.* Washington D.C: U.S. Government Printing Office.

Heun, K. 1900. Neue Methoden zur approximativen Integration der Differentialgleichungen einer unabhängigen Veränderlichen. *Zeitschrift Für Mathematik Und Physik* 45: 23–38.

INA. 1951. *Institute for Numerical Analysis (U.S.)—Problems for the Numerical Analysis of the Future*. Papers presented at the Symposia on Modern Calculating Machinery and Numerical Methods, held on July 29–31, 1948, at the University of California, Los Angeles. US Government Printing Office.

Israel, G., and A. Millán Gasca. 2009. *The World as a Mathematical Game. John von Neumann and Twentieth Century Science.* Basel: Birkhäuser.

Kirsch, R.A. 1998. SEAC and the start of image processing at the National Bureau of Standards. *IEEE Annals of the History of Computing* 20 (2): 7–13.

Kutta, W. 1901. Beitrag zur näherungsweisen Integration totaler Differentialgleichungen. *Zeitschrift für Mathematik und Physik* 46, 435–453.

Lee, J.A. 1995. *Computer Pioneers*. Retrieved from https://history.computer.org/pioneers/curtiss.html

McLachlan, R.I., K. Modin, H. Munthe-Kaas, and O. Verdier. 2017. Butcher series: A story of rooted trees and numerical methods for evolution equations. *Asia Pacific Mathematics Newsletter* 7 (1): 1–11.

Merson, R.H. 1957. An operational method for the study of integration processes. In: *Proceedings Symposium Data Processing. 1*, pp. 1–25. Salisbury, South Australia: Weapons Res.

Miklos, R (ed.). 2005. *John von Neumann: Selected Letters,* vol. 27. Providence, RI: American Mathematical Society/London Mathematical Society.

von Neumann, J. 1963. *Collected Works. Volume V: Design of Computers, Theory of Automata and Numerical Analysis,* ed, A.H. Taub. Oxford: Pergamon Press.

Nofre, D. 2014. Managing the technological edge: The UNESCO International Computation Centre and the limits to the transfer of computer technology, 1946–61. *Annals of Science* 71 (3): 410–431.

Nyström, E.J. 1925. Über die numerische Integration von Differentialgleichungen. *Acta Societatis Scientiarum Fennicae* 50: 1–54.

Pekeris, C.L., Z. Alterman, and L. Finkelstein. 1960. Solution of the Boltzmann-Hilbert integral equation, propagation of sound in a rarefied gas. In *Symposium on the Numerical Treatment of Ordinary Differential Equations, Integral and Integro-Differential Equations*, pp. 388–398. Basel: Birkhüaser.

Rabinowitz, P. 1952. The use of sub-routines on SEAC for numerical integrations of differential equations and for Gaussian quadrature. In *Proceedings of the 1952 ACM National Meeting*, pp. 88–89. Toronto

Ralston, A., and P. Rabinowitz. 1978. *First Course in Numerical Analysis*. New York: McGraw-Hill.

Redmond, K.C., and T. Smith. 1980. *Project Whirlwind. The History of a Pioneer Computer*. New York: Digital Press.

Reitwiesner, G.W. 1950. An ENIAC determination of pi and e to more than 2000 decimal places. *Mathematical Tables and Other Aids to Computation* 4 (29): 11–15.

Richey, M. 2010. The evolution of Markov Chain Monte Carlo methods. *The American Mathematical Monthly* 117: 383–413.

Runge, C. 1895. Über die numerische Auflösung von Differentialgleichungen. *Mathematische Annalen* 46: 167–178.

Small, J.J. 2013. *The Analogue Alternative: The Electronic Analogue Computer in Britain and the USA, 1930–1975.* London: Routledge.

Todd, J. 1990. The prehistory and early history of computation at the US National Bureau of Standards. In *A History of Scientific Computing*, ed. S.G. Nash and S. Nash, 251–268. Reading, MA: Addison-Wesley.

Turing, A.M. 1948. Rounding-off errors in matrix processes. *The Quarterly Journal of Mechanics and Applied Mathematics* 1 (1): 287–308.

Wilkes, M.V., D.J. Wheeler, and S. Gill. 1951. *The Preparation of Programs for an Electronic Digital Computer: With special reference to the EDSAC and the Use of a Library of Subroutines.* Cambridge, MA: Addison-Wesley.

Chapter 3
Integral Equations

Abstract From early on in his career, Pekeris became interested in integral equations, as they that arises in the context of seismology. Eventually he solved many important problems related with integral equations, often in collaboration with Zipora Alterman. They developed orginal methods for dealing with the Boltzmann-Hilbert equation for describing the behavior of gases.

Keywords WEIZAC · Chaim L. Pekeris · Zipora Alterman · Krzysztof Frankowski · Boltzmann-Hilbert Equation · Intergal Equations

An early important publication by Pekeris bore the title "A Pathological Case in the Numerical Solution of Integral Equations" (Pekeris 1940). He analyzed here an integral equation that arises in the context of seismology, in relation to the problem of the deformation affecting the surface of a uniform elastic half-space when a localized normal stress is suddenly applied to it. The equation in point is this:

$$f(u) = \frac{\sqrt{u + 1/3}}{(2u + 1)^2 - 4u\sqrt{u + 1/3}\sqrt{u + 1}} = \int\limits_{}^{\infty} \frac{H(v)dv}{(u + v)^{3/2}}. \tag{3.1}$$

The seismological problem at issue had originally been addressed by Horace Lamb (1849–1934) (Lamb 1904). Pekeris was well acquainted with Lamb's work on hydrodynamics and the elasticity of solid bodies, as will be discussed below in some detail. Equation (3.1) is an instance of what the classical mathematical theory originally developed by Ivar Fredholm (1866–1927), and further elaborated by David Hilbert (1862–1943), defines as an integral equation of the first kind:

$$f(u) = \int\limits_{0}^{\infty} H(v)K(v, u)dv.$$

Here, the function $f(u)$ and the kernel $K(v, u)$ are given, and $H(v)$ is the unknown function to be determined. The purely mathematical aspects of the theory of integral equations had received considerable interest over the preceding decades, and the

L. Corry and R. Leviathan, *Chaim L. Pekeris and the Art of Applying Mathematics with WEIZAC, 1955–1963*, SpringerBriefs in History of Science and Technology, https://doi.org/10.1007/978-3-031-27125-0_3

source of Pekeris's interest was related to their frequent appearance in significant physical situations. In such situations, the analytic form of the kernel $K(v, u)$ was often well known, whereas the values of $f(u)$ were obtained observationally at a discrete set of values of u. If, in addition, the observational errors associated with $f(u)$ are also known, then the question arises as to the degree of accuracy within which the unknown function $H(v)$ can be determined. If the equation is solved by numerical methods, moreover, one might expect that if an approximate value $f_1(u)$ is calculated for $f(u)$, then a trial value $H_1(v)$ would approximate the solution $H(v)$ by the same degree of accuracy that $f_1(u)$ approximates $f(u)$.

Pekeris's "pathological example" was intended to show that such an expectation does not hold true, and hence, more generally, that the uncertainty related to the use of integral equations in real, physical situations is one that requires focused attention and special care. Using two different trial functions, $f_1(u), f_2(u)$, Pekeris found two approximate solutions $H_1(v), H_2(v)$ to the equation, and he showed that: (1) Whereas $H_1(v)$, when plotted, is "totally inacceptable" as representing a real seismogram output, the resulting $f_1(u)$ reproduces the values of $f(u)$ to within 2%; (2) $H_2(v)$ is "physically wrong" by 25% and is "otherwise deficient" and yet it reproduces the values of $f(u)$ to within 0.1%. This is the "pathological behavior" that he wanted to call attention to.

Pekeris's interest in integral equations will resurface at various points in his career, particularly in relation to the Boltzmann equation, as we will see below. But it is interesting to stress that this interest actually started very early on. Indeed, the situation he investigated in 1940 had originally arisen in one of his earliest published articles dating from 1934 (Pekeris 1934). Still as a graduate student, with the encouragement of Rossby, Pekeris won a Guggenheim Fellowship that helped him study in Oslo with the eminent meteorologist Vilhelm Friman Bjerknes (1862–1951), as well as with the younger Svein Rosseland (1894–1985). Back then, Pekeris applied rather cumbersome numerical methods for solving an integral equation used to describe the distribution of ozone in the atmosphere. The focus of his article of 1934, however, was not the improvement of the numerical methods available or the creation of new ones or the properties of the equations, but rather the physical issues at stake, and in particular the methodological difficulties associated with the attempts to measure atmospheric magnitudes at low altitudes.

Pekeris's involvement with integral equations in general, with the Boltzmann equation in particular, and with the attempts to solve it by numerical methods with the help of WEIZAC arose from within the problem-setting just described, with its peculiar blend of mathematical and physical considerations. It is nonetheless important to stress the gradual, but clear, consolidation of Pekeris's self-assumed professional identity not so much as a physicist, or even specifically as a geophysicist or a meteorologist, but rather as "an applied mathematician."[1] Also, Alterman, who later became a leading figure in the department of geophysics at Tel Aviv University, subscribed to that kind of professional identity. In an article written in her honor by Keith Edward Bullen (1906–1976), of noted fame for his work on the deep structure

---

[1] See, e.g., Pekeris to von Neumann, Feb. 19, 1956 (CPA).

of the Earth's mantle and core (and about whom we say more below), he described her scientific persona in the following words:

> We both looked at Applied Mathematics as the art of deploying mathematical skills in contexts which lie outside the realm of pure mathematics. We shared the view that, with the applied mathematician, the context of application rather than the mathematics is the primary consideration, even though the mathematics brought to bear may sometimes need to be of a high order. Whereas a pure mathematician does mathematics for the sake of mathematics, the applied mathematician must have a considerable depth of understanding of the outside context. ... both of us had selected Geophysics as a principal context for much of our work. (Bullen 1976, pp. 17–18)

Before describing the joint efforts of Pekeris and Alterman around the Boltzmann equation in a more detailed manner, we devote the following section to a brief, but necessary, historical account of the background to the mathematical theory of integral equations.

3.1 From Fredholm to Hilbert and Beyond

Important contributions to the theory of integral equations had appeared at the end of the nineteenth century in the works of Giulio Ascoli (1843–1896), Vito Volterra (1860–1940), and Tullio Levi–Civita (1873–1941), before being approached in a more systematic way by Ivar Fredholm (Archibald and Tazzioli 2014). Hilbert's seminal contribution to the theory comprised reducing the solution of certain integral equations, and of some differential equations as well, to a problem in the theory of algebraic invariants. He treated the equations as limits of systems of an infinite number of linear equations, using infinite determinants to solve them and to prove important theorems of existence of solutions and convergence of series (Corry 2004, Chap. 5). Hilbert repeatedly stressed the analogy between algebra and analysis underlying his work, and several aspects of his efforts were devoted to establishing a stable bridge between the two disciplines by means of a theory of "analysis of infinite independent variables" (Hilbert 1909).

Hilbert was also clearly aware of the connection between his achievements in the theory of integral equations and their possible application to fundamental problems in physical domains and in the first place, from his own perspective, to the kinetic theory of gases (but also, beginning in the mid-1920s, the theory would become a central conceptual tool for quantum mechanics). Hilbert's first published work on a physical topic was devoted to the "mathematical foundations of the kinetic theory of gases," and the main challenge to be addressed in this context was the solution of the Boltzmann equation. This work appeared in 1912 as the last chapter of his treatise on the theory of linear integral equations (Hilbert 1912a, b). In this seminal work, Hilbert transformed the original equation into a more manageable one that came to be known as the Boltzmann-Hilbert equation and about which we say more below.

The most direct impact of Hilbert on the development of the theory of integral equations and its applications to physics came through the work of the Swedish physicist David Enskog (1884–1947), who had attended Hilbert's Göttingen lectures of 1911–12. Building on Hilbert's ideas, Enskog developed in (Enskog 1922) what came to constitute, together with the work of Sydney Chapman (1888–1970) (Chapman and Cowley 1939), the standard approach to the whole issue of transport phenomena in gases. Another standard point of reference for the theory, that Pekeris referred to in his works, was the work of Georg Hamel (1877–1954), one of Hilbert's early doctoral students (Hamel 1937).

The seminal works of Volterra and Fredholm on integral equations, as well as those influenced by Hilbert's algebraically oriented methods, led in the early decades of the twentieth century to the development of increasingly abstract approaches associated with functional analysis, including the theory of Banach spaces, spectral theory and normed rings (Bernkopf 1966; Siegmund-Schultze 2003, pp. 385–393). On the other hand, integral equations continued to be applied in various areas of physics, and in most cases, the aim was to obtain approximate numerical solutions. The example of Pekeris's meteorological work mentioned above was a case in point.

All the while, improved numerical methods were developed specifically for this purpose, where much attention was devoted to the concomitant question of error estimation. These methods comprised two main categories: (1) iterative procedures that led to the formulation of large systems of linear equations and (2) substitution of the original kernel $K(v, u)$ by a new one $K^*(v, u)$, that led to an equivalent equation, more amenable to numerical treatment (e.g., by expecting that $K^*(v, u)$ will give rise to a much smaller number of eigenvalues than $K(v, u)$) (Bückner 1952). All of these methods were developed prior to the rise of the electronic computer, but they became particularly useful thereafter as the implementation of powerful methods for inverting large matrices (as described above) and calculating their eigenvalues became feasible (Atkinson 2009; Lonseth 1954). Again, Pekeris's work on the Boltzmann equation in the mid-1950s arose as part of this trend.

3.2   Boltzmann-Hilbert Equation with WEIZAC

In his 1940 article, Pekeris did not indicate exactly what methods he had used for calculating the approximate solutions that he presented as examples of a "pathological case" for integral equations of the first kind. One can only speculate about the two main candidates for this role. One is the rule for numerical approximation of definite integrals, initially introduced by the British mathematician Thomas Simpson (1710–1761) and subsequently improved throughout the years. Pekeris explicitly mentioned this rule when explaining how integral equations of this kind had been solved in earlier articles that dealt with related issues (Pidduck 1917). The second one is the method developed by Nyström, already mentioned above in reference to the fifth-order RK algorithms. In 1930 Nyström had published an article on numerical solutions of precisely the kind of integral equations that Pekeris was now interested in (Nyström 1930). Nyström's approach was based on replacing the continuous integral by a weighted sum of n discrete intervals, by means of which the problem becomes that of solving a system of linear equations. It is unlikely that Pekeris had not been aware of Nyström's article.

Whatever the method, we do know something about the kind of machinery and automatic computing technologies available at the time that may have assisted the calculations both in 1940, when he arrived in Columbia, and previously at MIT. The leading figure in Columbia at the time was the astronomer Wallace John Eckert (1902–1971), who directed the Rutherford Laboratory. He had convinced IBM's president, Thomas J. Watson, to donate a newly developed IBM 601 calculating punch machine, with the help of which he developed important techniques for difficult mathematical tasks including numerical solutions of differential equations (Eckert 1940). Also, as already mentioned above, a relay computer of Bell Labs was made available to the team of the Division of War Research at Columbia University, which included Pekeris. In addition, over the 1940s many computation-intense mathematical projects in the USA, and in particular at Columbia, were carried out by groups of human (mainly women) computers working on specific tasks with the help of Marchant tabletop machines,[2] and in fact, Pekeris himself had relied there on the services of such human computers for his own research.

About fifteen years after having published his article on integral equations, WEIZAC was already operational in Rehovot, and the DAM at WIS was an already well-established center. Mathematical methods for geophysical research were a main topic of interest for Pekeris, as we will see below, and in this framework, he continued to devote much attention to the analytic treatment of integral equations (Pekeris 1955, 1956a). Naturally, however, with the availability of WEIZAC, the numerical treatment of the Boltzmann equation and of questions related to the behavior of rarefied gases also became increasingly high on his agenda. Alterman was also involved, between 1955 and 1957, in all the work that led to the publication of numerical values for the coefficients of viscosity and heat conductions in gases. Her reports to Pekeris provide detailed information on progress at both the level of working out the mathematical details of the solutions and of coding the machine. Thus, for instance, in April 1957 Alterman reported on the lower accuracy attained in calculations performed with the method of "integer and fraction," as opposed to that of double-precision floating point (about which, more is said below in Sect. 15.2). She also explained to Pekeris the basic structure of the program used for the numerical solution of the equation in terms of a block diagram displaying the main flow and the various subroutines, using the diagram reproduced in Fig. 3.1.

The meaning of the blocks and the flow is explained in Fig. 3.2.

The programs run with WEIZAC calculated the desired numerical values by ingeniously transforming the given integral equation into an ODE, to which well-known numerical methods could be applied (Pekeris 1956b; Pekeris and Alterman 1957). Some of these methods had been introduced much earlier, but they were never fully exploited in relation to the Boltzmann-Hilbert equation. In order to explain

---

[2] "Columbia University Computing History; A Chronology of Computing at Columbia University", http://www.columbia.edu/cu/computinghistory/#year1940.

Fig. 3.1 Alterman's original block diagram of the program used for numerically solving the Boltzmann equation ("Monthly report for April 1957," in "Progress Reports for the Years 1956–57," p. 4 (CPA)). See Fig. 3.2 for an English translation.

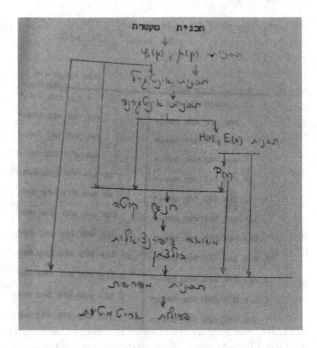

the novelty involved in the approach followed by Pekeris and Alterman, it is first necessary to recall some of the details of previous work done on the equation.

The behavior of a perfect gas, as already indicated, was typically characterized by assuming that the state of any of its molecules, considered to be rigid spheres, is independent of all the others, except at the instant when they collide. The classical problems of gas theory involve the determination of the transport coefficients μ, λ, and D of viscosity, heat conduction, and diffusion, respectively. In a classical work of 1872,[3] Boltzmann specified the position of a molecule at a given time t, in space $\mathbf{r}$ and in the velocity-space $\mathbf{c}$, by means of a distribution function $f(\mathbf{c}, \mathbf{r}, t)$. If we let $\mathfrak{D}f$ be:

$$\mathfrak{D}f \equiv \frac{\partial f}{\partial t} + \mathbf{c} \cdot \frac{\partial f}{\partial \mathbf{r}} + \mathbf{F} \cdot \frac{\partial f}{\partial \mathbf{c}}, \tag{3.2}$$

then, function f is controlled by the Boltzmann equation, which we formulate here, following Pekeris, as:

$$\mathfrak{D}f = \frac{\sigma^2}{2} \iint \left(f' f_1' - f f_1 \right) |\mathbf{g} \cdot \mathbf{k}| d\mathbf{k} d\mathbf{c}_1. \tag{3.3}$$

Here $\mathbf{F}$ denotes the force per unit mass acting on the molecules; σ is the diameter of the molecules; and $\mathbf{k}$ is a unit vector in the direction of the line of the centers at

---

[3] For historical details see (Cercignani 1998), esp. Chap. 4.

Fig. 3.2 Alterman's block
diagram

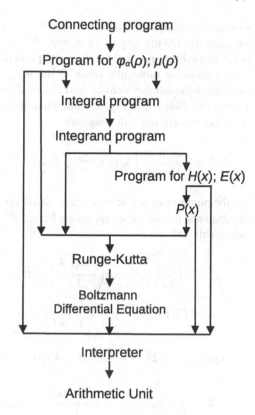

Connecting program

Program for $\varphi_\alpha(\rho)$; $\mu(\rho)$

Integral program

Integrand program

Program for $H(x)$; $E(x)$

$P(x)$

Runge-Kutta

Boltzmann
Differential Equation

Interpreter

Arithmetic Unit

the time of the collision. The velocities of two colliding particles are, respectively,
(i) after collision: c, c_1 and (ii) before collision: c', c'_1. In addition, g denotes the
difference $c' - c$. Accordingly, we have:

$$f = f(c, r, t); \quad f_1 = f(c_1, r, t); \quad f = f(c', r, t); \quad f_1 = f(c'_1, r, t)$$
$$\text{and } c' = c + (g \cdot k)k; \quad c'_1 = c_1 - (g \cdot k)k.$$

In addition, for any property $\phi\,(c, r, t)$ of the flow, the average $\bar{\phi}(r, t)$ taken over
the entire velocity-space c is given by:

$$\bar{\phi}(r, t) = \int \phi(c, r, t) f(c, r, t) dc.$$

Boltzmann had not been able to determine the transport coefficients because of
the difficulty involved in solving the equation, and it was here that Hilbert entered
the stage in 1912. The confluence of a deep physical problem with a truly challenging mathematical task was absolutely appealing for Hilbert. Increased appeal
derived from the fact that—in Hilbert's understanding—the Boltzmann equation

was an "intrinsic" integral equation, namely one that did not originate in a differential equation. On the way to a solution, Hilbert simplified the equation significantly as he carried through, by an elaborate procedure, integration over **k**. Thus, if we let m denote the molecular mass, T the temperature of the gas, k the Boltzmann constant, n the number density, and φ a small magnitude, then, Hilbert transformed the original Boltzmann Eq. 3.3 into what became known as the Boltzmann-Hilbert equation, namely the following one:

$$\mathfrak{D}f^0 = -\frac{\sigma^2 n^2 m}{2\pi kT}\left\{M(p)e^{-2p^2}\varphi + \frac{1}{\pi}e^{-p^2}\int \varphi_1 e^{-p_1^2}\left(R - \frac{2}{R}e^{w^2}\right)d\mathbf{p}_1\right\} \quad (3.4)$$

For the purposes of our account here, we do not need to go into all the details of this equation and the variables appearing here. Still, for the sake of completeness we do indicate the following:

$$f^0 = n\left(\frac{m}{2\pi kT}\right)^{\frac{3}{2}}\exp\left(\frac{-mC^2}{2kT}\right), \quad C = c - c_0,$$

$$f = f^0(1 + \varphi), \quad \mathbf{p} = C\sqrt{\frac{m}{2kT}},$$

$$M(p) = 1 + \left(2p + \frac{1}{p}\right)P(p), \quad P(p) = e^{p^2}\int_0^p e^{-x^2}dx$$

$$R = |\mathbf{p} - \mathbf{p}_1|, \quad \omega = \frac{pp_1\sin\theta'}{R} \quad (\theta' \text{ denotes the angle between } \mathbf{p}, \mathbf{p}_1). \quad (3.5)$$

Notice that in this integral equation, φ is the unknown function. The meaning of the conditions stated here is that the deviation f^0 can be treated as a first-order perturbation and hence the equation can be linearized.

Now, starting from Eq. (3.4), Pekeris and Alterman proceeded to determine the values of λ, D, and μ.[4] Each of these transport coefficients involves some specific conditions that lead to a special case of the integral equation on which they focused, one at a time. Thus, for instance, in the case of μ, the coefficients of viscosity, n and T, are taken to be constant, and the value of the coefficient turns out to be given by:

$$\mu = \frac{8}{15\pi\sigma^2}\sqrt{mkT}\int_0^\infty e^{-p^2}b(p)p^6 dp, \quad (3.6)$$

---

[4] Although we are referring here only to the work of Pekeris and Alterman as it appears in the published articles, it is worth pointing out that Alterman's research reports for the years 1956–57 also contain detailed information about some of the issues encountered, both at the theoretical and the computational level, in their work on these issues. Alterman informed Pekeris every month about progress attained in each project and continually stressed that each new routine programmed for WEIZAC was also checked manually.

where $b(p)$ is, in turn, a solution of the equation:

$$M(p)e^{-p^2}b(p)p_xp_y + \frac{1}{\pi} \int b(p_1)e^{-p_1^2}p_{x1}p_{y1}\left(R - \frac{2}{R}e^{w^2}\right)d\mathbf{p}_1 = p_xp_y. \quad (3.7)$$

The situation involved here may be understood in terms related to the Fredholm equation of the *second* kind, namely an equation of the type:

$$f(u) = g(u) + \rho \int_a^b f(v)K(v, u)dv,$$

where $g(u)$ and the kernel $K(v, u)$ are given, and $f(v)$ is the unknown function to be determined, like $b(p)$ in the present case.

The equations arising for each of the three transport coefficients λ, D, and μ (which we do not copy here for the sake of brevity) are of the same kind and, indeed, the kernel $\left(R - \frac{2}{R}e^{\omega^2}\right)$ is the same one for the three equations. Pekeris and Alterman realized that a very convenient way to numerically approximate the values of the coefficients would be based on a suitable expansion of the kernel. They chose to do so with the help of the "spherical harmonics" of the angle θ', which is the angle between $\mathbf{p}$ and $\mathbf{p}_1$. There was, to be sure, a good mathematical reason for this choice, namely that in each case the equation considered involved only a *single spherical harmonic* in $\mathbf{p}$, which would help simplifying the approximation procedure. In the case of viscosity, for instance, one sees in Eq. (3.7) that only the spherical harmonic of order 2, p_xp_y (p_k being components of $\mathbf{p}$) is actually involved. Likewise, for the evaluation of λ and D they showed that the coefficients of the first and second harmonic would suffice for the expansion. More specifically, the kernel in the case of viscosity can be written as follows:

$$\left(R - \frac{2}{R}e^{\omega^2}\right) = \sum_{n=0}^{\infty}\left(n + \frac{1}{2}\right)A_n(p, p_1)P_n(\cos\theta'). \quad (3.8)$$

By doing so, a somewhat lengthy series of successive simplifications of the original integral equation, combined with exacting manipulations of variables, yield explicit expressions for the coefficients $A_n(p, p_1)$, which are symmetrical in the arguments p, p_1. It is at this point that the crucial step in the approach followed by Pekeris and Alterman comes into play, as they were able to transform, for each of the transport coefficients, the corresponding Boltzmann-Hilbert equation into an ODE, which in turn could be numerically approximated using the techniques that they were well acquainted with. Assisted by WEIZAC, they could surely undertake the calculations with great speed and accuracy.

Completing the necessary tasks required extremely laborious calculations, which we skip here. Suffice it to say that in 1955 Pekeris treated the case of self-diffusion and obtained a differential equation of the second order. He provided explicit expressions for A_0, A_1, and A_2 and, later, he also calculated values of A_n for up to $n = 20$. WEIZAC became operational in September of 1955, which means that these calculations were still done with the help of desktop manual devices. In 1957, working together with Alterman, Pekeris approached the problems of heat conduction and viscosity, which yield differential equations of the fourth order for the respective distribution functions. In all three cases, the solutions of the differential equations were calculated with WEIZAC,[5] yielding values for the coefficients, which were in very good agreement with the values obtained by the Chapman-Enskog method. Still Pekeris and Alterman emphasized that the linearized approximation became relatively poorer in the outer regions of the momentum space.

A curious detail pointed out by Pekeris was that, when studying Boltzmann's complete works after having completed his own research, he had come across a rarely cited text (probably also unknown to Hilbert) where Boltzmann had derived linear differential equations from the integral one, by very laborious methods. In the case of self-diffusion, this was a second-order differential equation, and in the case of viscosity, it was fourth-order one.[6] Pekeris also identified some errors that crept into the last stages of Boltzmann's derivation, which, in any case, he did not integrate. As Pekeris pointed out, it is remarkable that Boltzmann referred to this derivation in his influential chapter on the kinetic theory, written together with Josef Nabl (1876–1953) for Felix Klein's influential *Encyklopädie der mathematischen Wissenschaften* (Gispert 1999). Boltzmann had remarked there, however, that attempts to solve the typical cases of viscosity and diffusion on the assumption that the molecules are elastic spheres led "to very complicated series-expansions which are hardly suited for practical computation, or to a differential equation of the fourth order whose coefficients are transcendental functions" (Boltzmann and Nabl 1904, p. 541).

Pekeris and Alterman concluded their article by stating that if Boltzmann had carried out the numerical integration, his efforts, spanning hundreds of printed pages full with dense calculation, "would have borne fruit" and he could have successfully obtained a value for the coefficient of viscosity. At the same time, they pointed out that, to the best of their knowledge, the differential equation derived and solved in their papers was totally new. This was no doubt true, but no less true was the fact (curiously not stressed by them here) that without the availability of a powerful electronic computer such as WEIZAC and without the programming skills that people like Alterman had started to develop at the time, the numerical values obtained with

---

[5] It is remarkable that, in spite of his considerable experience with programmable machines going back to his wartime activities at Columbia and MIT, Pekeris continued to express a child-like enthusiasm for this kind of activity. In a letter to Estrin, where he informed him about current activities in seismology, Pekeris also wrote: "I have at last sat down and learned coding and yesterday I finished my second programme. It is quite thrilling to see the beast turn out answers in minutes where the hand computers spend days, and the computer also does it with double-precision floating." He also added, however, that he was starting now the coding of a third program, after which he would "feel justified in quitting the coding business and have my assistants do the coding for me, as in the past." Pekeris to Estrin, June 3, 1957 (CPA). See also footnote 3 above.

[6] Pekeris gives a reference to (Boltzmann 1909), Vol. 2, p. 545 and Vol. 3, p. 48.

the help of the equations thus derived may not have been obtained at all, and certainly not that fast and not to that degree of certitude and accuracy.

3.3 Further Work on Integral Equations

To conclude this section, it is relevant to mention some additional research undertaken by Pekeris and his collaborators in numerically solving integral equations with the help of WEIZAC (mainly, but not exclusively, in relation to the Boltzmann-Hilbert equation), and to briefly comment on the impact of these works.

In 1962, Pekeris published together with Alterman and with Frankowski (who was to complete his Ph.D. in 1964 under Pekeris working on these topics) further calculations related with the equation (Alterman et al. 1962). This time they calculated over 550 eigenvalues for an ideal gas and for a rigid-sphere gas. Their calculations with WEIZAC were remarkably accurate, relative to anything previously known. The values were first computed in double-precision arithmetic and then double-checked with triple-precision floating arithmetic. These results agreed to 11 decimal figures and in some more specific cases up to 18 decimals. That year the three published yet another article (Pekeris et al. 1962), this time together with Leib Finkelstein (who also completed a Ph.D. under Pekeris). Here, the physical context of the investigation was that of the propagation of sound in a gas of rigid-sphere molecules. The basic approach relied once again on reducing the integral equation into an ODE. Assisting themselves with WEIZAC, they were able to solve determinants of the order of 105 in order to obtain the desired eigenvalues.

Of particular interest for us here is that the complex details of the calculation are explicitly described in the article, and they definitely help to shed light on the intrinsic difficulties of programming WEIZAC to calculate the numerical values. The kernels $A_n(p, p_1)$ of Eq. (3.8) are represented by daunting expressions, such as the following two, which we reproduce here in Fig. 3.3, as they appear from the original article of the four authors.

In this way, the integral equation is transformed into a set of $(2k + 2)$ simultaneous first-order linear differential equations, with $k = [n/2]$. The equations were then numerically integrated on WEIZAC by standard RK-type methods.

The contributions of Pekeris and his collaborators to the further development of methods of solution for the Boltzmann equation became central to this important field of enquiry. Their works continued to be cited for decades to come, often as a substantial starting point for many important articles.[7]

---

[7] See, e.g., (Siewert 2003). Google Scholar counts, up to 2010, more than 130 citations for (Pekeris and Alterman 1957).

PROPAGATION OF SOUND IN GAS OF RIGID SPHERES 1611

TABLE II. The kernels $A_n(p, p_1)$ of Eq. (14) for $p_1 < p$.

$$(pp_1)^3 A_3(p, p_1) = P(p_1)[150 + 30p^2 + (-120 - 21p^2)p_1^2 + (45 + 5p^2)p_1^4 - 10p_1^6]$$
$$+ (-150 - 30p^2)p_1 + (20 + p^2)p_1^3 - 5p_1^5 - \frac{2}{35}p^2 p_1^7 + \frac{2}{63}p_1^9;$$

$$(pp_1)^5 A_4(p, p_1) = P(p_1)\left[-\frac{3675}{2} - \frac{1575}{4}p^2 - \frac{105}{8}p^4 + \left(\frac{5775}{4} + \frac{1125}{4}p^2 + \frac{15}{2}p^4\right)p_1^2 + \left(-\frac{4305}{8} - 90p^2 - \frac{3}{2}p^4\right)p_1^4 + \left(\frac{245}{2} + 15p^2\right)p_1^6\right.$$
$$\left.- \frac{35}{2}p_1^8\right] + \left(\frac{3675}{2} + \frac{1575}{4}p^2 + \frac{105}{8}p^4\right)p_1 + \left(-\frac{875}{4} - \frac{75}{4}p^2 - \frac{5}{4}p^4\right)p_1^3 + \left(\frac{525}{8} + \frac{15}{2}p^2\right)p_1^5 - \frac{35}{4}p_1^7 - \frac{2}{63}p^2 p_1^9 + \frac{2}{99}p_1^{11};$$

$$(pp_1)^6 A_5(p, p_1) = P(p_1)\left[\frac{59535}{2} + 6615p^2 + \frac{2835}{8}p^4 + \left(-\frac{46305}{2} - \frac{9555}{2}p^2 - \frac{1785}{8}p^4\right)p_1^2 + \left(\frac{68985}{8} + \frac{6405}{4}p^2 + 60p^4\right)p_1^4\right.$$
$$+ \left(-\frac{16065}{8} - 315p^2 - \frac{15}{2}p^4\right)p_1^6 + \left(315 + 35p^2\right)p_1^8 - \frac{63}{2}p_1^{10}\right] + \left(-\frac{59535}{2} - 6615p^2 - \frac{2835}{8}p^4\right)p_1$$
$$+ \left(\frac{6615}{2} + \frac{735}{2}p^2 + \frac{105}{8}p^4\right)p_1^3 + \left(-\frac{9009}{8} - \frac{721}{4}p^2 - \frac{23}{4}p^4\right)p_1^5 + \left(\frac{1323}{8} + \frac{35}{2}p^2\right)p_1^7 - \frac{63}{4}p_1^9 - \frac{2}{99}p^2 p_1^{11} + \frac{2}{143}p_1^{13};$$

$$(pp_1)^7 A_6(p, p_1) = P(p_1)\left[-\frac{2401245}{4} - \frac{1091475}{8}p^2 - \frac{72765}{8}p^4 - \frac{3465}{32}p^6 + \left(\frac{3711015}{8} + 99225p^2 + \frac{191835}{32}p^4 + \frac{945}{16}p^6\right)p_1^2\right.$$
$$+ \left(-\frac{1382535}{8} - \frac{1091475}{32}p^2 - \frac{28665}{16}p^2 - \frac{105}{8}p^4\right)p_1^4 + \left(\frac{1306305}{32} + \frac{114975}{16}p^2 + \frac{2415}{8}p^4 + \frac{5}{4}p^6\right)p_1^6$$
$$+ \left(-\frac{107415}{16} - \frac{7875}{8}p^2 - \frac{105}{4}p^4\right)p_1^8 + \left(\frac{6237}{8} + \frac{315}{4}p^2\right)p_1^{10} - \frac{231}{4}p_1^{12}\right]$$
$$+ \left(\frac{2401245}{4} + \frac{1091475}{8}p^2 + \frac{72765}{8}p^4 + \frac{3465}{32}p^6\right)p_1 + \left(-\frac{509355}{8} - \frac{33075}{4}p^2 - \frac{2205}{32}p^4 + \frac{105}{8}p^6\right)p_1^3$$
$$+ \left(\frac{189189}{8} + \frac{138915}{32}p^2 + \frac{441}{2}p^4 + \frac{21}{8}p^6\right)p_1^5 + \left(-\frac{114345}{32} - \frac{4095}{8}p^2 - \frac{105}{8}p^4\right)p_1^7 + \left(\frac{1617}{4} + \frac{315}{8}p^2\right)p_1^9 - \frac{231}{8}p_1^{11} - \frac{2}{143}p^2 p_1^{13} + \frac{2}{195}p_1^{15}.$$

Fig. 3.3 Two explicit expressions for the kernels $A_n(p, p_1)$. Reproduced from (Pekeris et al. 1962), with permission of AIP Publishing

TABLE III. The kernels $A_n(p, p_1)$ of Eq. (14) for $p_1 < p$.

$$(pp_1)^9 A_7(p, p_1) = P(p_1)\left[\frac{579972915}{4} p^2 + \frac{13378365}{4} p^2 + \frac{2027025}{8} p^4 + \frac{45045}{8} p^6 + \left(-\frac{22297275}{2} - \frac{19552995}{8} p^2 - \frac{343035}{2} p^4\right.\right.$$

$$\left.-\frac{107415}{32} p^6\right)p_1^2 + \left(\frac{16621605}{8} + \frac{6829515}{8} p^2 + \frac{1732185}{32} p^4 + \frac{14175}{16} p^6\right)p_1^4 + \left(-\frac{5977125}{32} - \frac{164115}{16} p^4 - \frac{1015}{8} p^6\right)p_1^6$$

$$+ \left(\frac{5360355}{32} + \frac{446985}{16} p^2 + \frac{9765}{8} p^4 + \frac{35}{4} p^6\right)p_1^8 + \left(-\frac{333333}{16} - \frac{22869}{8} p^2 - \frac{315}{4} p^4\right)p_1^{10} + \left(\frac{15015}{8} + \frac{693}{4} p^2\right)p_1^{12}$$

$$-\frac{429}{4} p_1^{14}\right] + \left(-\frac{57972915}{4} - \frac{13378365}{4} p^2 - \frac{2027025}{8} p^4 - \frac{45045}{8} p^6\right)p_1 + \left(\frac{1486485} + \frac{1715175}{8} p^2 + \frac{10395}{4} p^4 - \frac{12705}{32} p^6\right)p_1^3$$

$$+ \left(-\frac{2351349}{4} - \frac{929313}{8} p^2 - \frac{235305}{32} p^4 - \frac{1197}{8} p^6\right)p_1^5 + \left(\frac{178893}{2} + \frac{467181}{32} p^2 + 603p^4 + \frac{19}{8} p^6\right)p_1^7$$

$$+ \left(-\frac{349635}{32} - \frac{11781}{8} p^2 - \frac{315}{8} p^4\right)p_1^9 + \left(\frac{3861}{4} + \frac{693}{8} p^2\right)p_1^{11} - \frac{429}{8} p_1^{13} - \frac{2}{195} p_1^{18} + \frac{2}{255} p_1^{17};$$

$$(pp_1)^9 A_8(p, p_1) = P(p_1)\left[-\frac{13043905875}{32} - \frac{6087156075}{32} p^2 - \frac{1003377375}{128} p^4 - \frac{30405375}{128} p^6 - \frac{675675}{512} p^8 + \left(\frac{2000065675}{64}\right.\right.$$

$$+ \frac{4463914455}{64} p^2 + \frac{6912155255}{128} p^4 + \frac{19054035}{128} p^6 + \frac{45045}{64} p^8\right)p_1^2 + \left(-\frac{14916876975}{128} - \frac{3152834685}{128} p^2 - \frac{451091025}{256} p^4\right.$$

$$\left.-\frac{343035}{8} p^6 - \frac{10395}{64} p^8\right)p_1^4 + \left(\frac{3581753175}{128} + \frac{706620915}{128} p^2 + \frac{11382525}{32} p^4 + \frac{116235}{16} p^6 + \frac{315}{16} p^8\right)p_1^6$$

$$+ \left(-\frac{2465538075}{512} - \frac{3468465}{4} p^2 + \frac{1541925}{32} p^4 - \frac{5985}{8} p^6 - \frac{35}{32} p^8\right)p_1^8 + \left(\frac{39864825}{64} + \frac{1376575}{16} p^2 + \frac{17325}{4} p^4 + \frac{315}{8} p^6\right)p_1^{10}$$

$$+ \left(-\frac{3918915}{64} - \frac{63063}{8} p^2 - \frac{3465}{16} p^4\right)p_1^{12} + \left(\frac{70785}{16} + \frac{3003}{2} p^2\right)p_1^{14} - \frac{6435}{32} p_1^{16}\right] + \left[\frac{1304390875}{32} + \frac{6087156075}{64}\right.$$

$$+ \frac{1003377375}{128} p^4 + \frac{30405375}{128} p^6 + \frac{675675}{512} p^8 + \left(-\frac{2608781175}{64} - \frac{405810405}{64} p^2 - \frac{22297275}{128} p^4 + \frac{1216215}{128} p^6 + \frac{45045}{256} p^8\right)p_1^3$$

$$+ \left(\frac{2162835675}{128} + \frac{447431985}{128} p^2 + \frac{64604925}{256} p^4 + \frac{446085}{64} p^6 + \frac{5775}{128} p^8\right)p_1^5 + \left(-\frac{328899285}{128} - \frac{57918861}{128} p^2 - \frac{3042765}{128} p^4\right.$$

$$\left.-\frac{8829}{32} p^6 + \frac{93}{64} p^8\right)p_1^7 + \left(\frac{168243075}{512} + \frac{3288285}{64} p^2 + \frac{142065}{64} p^4 + \frac{315}{16} p^6\right)p_1^9 + \left(-\frac{8140275}{256} - \frac{129129}{32} p^2 - \frac{3465}{32} p^4\right)p_1^{11}$$

$$+ \left(\frac{289575}{128} + \frac{3003}{16} p^2\right)p_1^{13} - \frac{6435}{64} p_1^{15} - \frac{2}{255} p_1^{17}p_1^{17} + \frac{2}{323} p_1^{17}.$$

Fig. 3.3 (continued)

References

Alterman, Z., K. Frankowski, and C.L. Pekeris. 1962. Eigenvalues and eigenfunctions of the linearized Boltzmann collision operator for a Maxwell gas and for a gas of rigid spheres. *The Astrophysical Journal* 7: 291.

Archibald, T., and R. Tazzioli. 2014. Integral equations between theory and practice: The cases of Italy and France to 1920. *Archive for History of Exact Sciences* 68 (5): 547–597.

Atkinson, K. 2009. A personal perspective on the history of the numerical analysis of Fredholm integral equations of the second kind. In *The Birth of Numerical Analysis* 53–72. World Scientific.

Bernkopf, M. 1966. The development of function spaces with particular reference to their origins in integral equation theory. *Archive for History of Exact Sciences* 3: 1–96.

Boltzmann, L., and Nabl, J. 1904. Kinetische Theorie der Materie. In *Encyclopädie der Mathematischen Wissenschaften*, Vol. V, 493–557. Leipzig: Teubner Verlag.

Boltzmann, L. 1909. *Wissenschaftliche Abhandlungen, 3 Vols.* (F. Hasenöhrl, Ed.) Leipzig: Barth.

Bückner, H. 1952. *Die Praktische Behandlung von Integral-Gleichungen.* Berlin: Springer.

Bullen, K.E. 1976. Free Earth oscillations and the structure of the Earth's interior. *New Zealand Mathematical Chronicle* 5: 17–45.

Cercignani, C. 1998. *Ludwig Boltzmann. The Man Who Trusted Atoms.* Oxford: Oxford University Press.

Chapman, S., and T.G. Cowley. 1939. *The Mathematical Theory of Non-Uniform Gases; An Account of the Kinetic Theory of Viscosity, Thermal Conduction and Diffusion in Gases.* Cambridge: Cambridge University Press.

Corry, L. 2004. *David Hilbert and the Axiomatization of Physics (1898–1918). From Grundlagen der Geometrie to Grundlagen der Physik.* Dordrecht: Springer.

Eckert, W. J. 1940. *Punched Card Methods in Scientific Computation.* Thomas J. Watson Astronomical Computing Bureau, Columbia University.

Enskog, D. 1922. Kinetische Theorie der Wärmeleitung, Reibung und Selbstdiffusion in gewissen verdichteten Gasen und Flüssigkeiten. *Kungliga Svenska Vetenskapsakademiens Handlingar* 63 (4): 1–44.

Gispert, H. 1999. Les débuts de l'histoire des mathématiques sur les scènes internationales et le cas de l'entreprise encyclopédique de Felix Klein et Jules Molk. *Historia Mathematica* 26 (4): 344–360.

Hamel, G. 1937. *Integralgleichungen.* Berlin: Springer.

Hilbert, D. 1909. Wesen und Ziele einer Analysis der unendlichvielen unabhangigen Variablen. *Rendiconti Del Circolo Matematico Di Palermo* 27 (1): 59–74.

Hilbert, D. 1912a. *Grundzuge Einer Allgemeinen Theorie der Linearen Integralgleichungen.* Leipzig: Teubner.

Hilbert, D. 1912b. Begründung der kinetischen Gastheorie. *Mathematische Annalen* 72: 562–577.

Lamb, H. 1904. Waves, surface, und tremors, in an elastic solid. *Philosophical Transactions of the Royal Society, A* 203: 1–42.

Lonseth, A.T. 1954. Approximate solutions of Fredholm-type integral equations. *Bulletin of the American Mathematical Society* 60 (5): 415–430.

Nyström, E.J. 1930. Über die praktische Auflösung von Integralgleichungen mit Anwendungen auf Randwertaufgaben. *Acta Mathematica* 54: 185–204.

Pekeris, C.L. 1934. On the interpretation of atmospheric ozone measurements. *Gerlands Beiträge Zur Geophysik* 41: 192–303.

Pekeris, C.L. 1940. A pathological case in the numerical solution of integral equations. *Proceedings of the National Academy of Sciences of the United States of America* 26 (6): 433.

Pekeris, C.L. 1955. The seismic surface pulse. *Proceedings of the National Academy of Sciences of the United States of America* 41 (7): 469–480.

Pekeris, C.L. 1956a. Solution of an integral equation occurring in impulsive wave propagation problems. *Proceedings of the National Academy of Sciences of the United States of America* 42 (7): 439–443.

Pekeris, C.L. 1956b. Solution of the Boltzmann-Hilbert integral equation. *Proceedings of the National Academy of Sciences of the United States of America* 41 (9): 661–669.

Pekeris, C.L., and Alterman, Z. 1957. Solution of the Boltzmann-Hilbert integral equation II. The coefficients of viscosity and heat conduction. *Proceedings of the National Academy of Sciences of the United States of America* 43 (11): 998–1007.

Pekeris, C.L., Z. Alterman, L. Finkelstein, and K. Frankowski. 1962. Propagation of sound in a gas of rigid spheres. *The Physics of Fluids 5*, 1608–1610. Retrieved from url = {https://doi.org/10.1063/1.1706572}.

Pidduck, F.B. 1917. The kinetic theory of the motion of ions in gases. *Proceedings of the London Mathematical Society* 15(1): 89–127.

Siegmund-Schultze, R. 2003. The origins of functional analysis. In *A History of Analysis* ed. Jahnke, H.-N. 385–408. Providence, RI: AMS/LMS.

Siewert, C.E. 2003. The linearized Boltzmann equation: Concise and accurate solutions to basic flow problems. *Zeitschrift Für Angewandte Mathematik Und Physik* 54: 273–303.

Chapter 4
Oscillations of the Earth

Abstract Even before his arrival in Israel, Pekeris was involved in efforts to organize a Scientific Advisory Committee for the establishment of a seismological observatory in Palestine. Between 1951 and 1955 a major share of the activities of DAM were related to geophysical prospecting. The geological survey started in the summer of 1950, and on September 23, 1955, oil was found in the Heletz field. Pekeris's collaboration with Zipora Alterman started around this time. They used WEIZAC to perform accurate calculations that afforded the first full proof of the existence of "free vibrations of the Earth." The calculated magnitudes of these vibrations were identical to those predicted by the existing theory.

Keywords WEIZAC · Chaim L. Pekeris · Zipora Alterman · Hans Jarosch · Israel Geological Survey · Caltech Seismo Lab · Beno Gutenberg · Hugo Benioff · Free oscillations of the earth · Earthquake modelling · Oil exploration

On May 22, 1960, a devastating earthquake shook the mainland of Chile. This was the most powerful earthquake ever recorded, leaving a frightening death toll and monetary losses that could never be established with certainty. At the scientific level, however, this frightful seismic event provided the first full proof of the existence of "free vibrations of the Earth." Moreover, the magnitude and characteristics of these vibrations were identical to those predicted by the theory and accurately calculated with the assistance of WEIZAC in a paper of 1959 published by Pekeris, together with Alterman et al. (1959) (henceforth jointly referred to as AJP). Like Pekeris's paper on the ground state of helium, discussed in detail below, this highly influential paper became a landmark in its field and it continues to be cited to this day.[1]

This remarkable scientific achievement embodies a synthesis of Pekeris's agenda as both a researcher and an institution builder. Indeed, when he joined WIS in 1949, he proposed a few projects to be carried out in the DAM. Two of the projects were for him a must: the building of an electronic computer and the pursuit of geophysics as a central discipline in the department.[2] At the time, Pekeris already

---

[1] Google Scholar indicates more than 230 citations after 2000.

[2] Bergman to Pekeris, Jul. 30, 1947 (CPA).

enjoyed a well-established international reputation in both theoretical and computational geophysics. As already indicated, his interest in the theoretical sides of the discipline started early on in his academic career, just after completing his Ph.D. at MIT in 1933. Back in 1934, Pekeris had published his first paper in geophysics, dealing with an inverse boundary problem in seismology (Pekeris 1934).[3] In 1936, he was promoted to associate geophysicist at MIT. He then started to deepen his knowledge in the theory of propagation of pulses and waves, publishing in 1941 a brief abstract on the subject in the *Transactions of the American Geophysical Union* (Pekeris 1941).

Geophysics, to be sure, was among the disciplines for which the expected gains of using electronic computers were acknowledged from very early on, and one should not be surprised to hear John von Neumann making explicit statements to this effect. In 1954, he gave a talk at Columbia, on the occasion of the dedication of IBM's Naval Ordnance Research Calculator (NORC), one of the most powerful electronic computers at the time.[4] Von Neumann spoke about the outstanding uses he envisaged for NORC in geophysics—alongside meteorology and the study of tides—and stated that the field presented unique challenges and possibilities for computer application because of the vast amounts of data to be processed and the spatial and temporal largeness of the multidimensional problems that it investigates (MacDonald 1962).

Pekeris became a central figure in this trend of applying electronic computers to geophysics. In 1956, he was the keynote speaker in a conference on theoretical geophysics jointly sponsored by the National Science Foundation and the Carnegie Institution of Washington. He was also invited as a distinguished participant in the first meeting of the Committee on Geophysical Theory and Computers (in 1964 in Moscow and Leningrad) and became the host for the second meeting, held the following year in Rehovot (Freeman 2004). As we saw above, Pekeris's interest in numerical solutions of complex mathematical problems arising from physical situations started very early on and with no direct connection to the possible development of automatic electronic calculation devices. When the latter started to become more readily available and central to numerical analysis, and as Pekeris became their foremost promoter at WIS, two lines of interest found a natural point of confluence that led to important results. This was the case also in the area of geophysics and seismography, where he also had the fortune to work in association with brilliant younger researchers, foremost of which was Zipora Alterman. We proceed now to discuss their contributions to seismology using WEIZAC, after briefly describing the historical background of the discipline in general.

4.1 From Classical Physics to Caltech's Seismographs

The study of free oscillations arises in the study of the Earth's dynamic response to external or internal forces. These are standing waves that propagate along the

---

[3] Inverse boundary value problems involve the calculation of the values of a function that satisfy certain boundary conditions, as opposed to determining the boundary conditions for a given function.

[4] See http://www.columbia.edu/cu/computinghistory/norc.html.

surface of the Earth, and their peaks can be observed in the frequency spectrum of large earthquakes. Alterman likened the free oscillations to the vibrations of a drum: "when a drum is hit, it vibrates. In a similar way the earthquake 'hits' the Earth and oscillations result" (Alterman et al. 1974, p. 409). Pekeris explained their meaning by comparison with the tone of a bell. "If you compare the tone of bells of different sizes—he wrote in 1958 on the occasion of receiving the Weizmann Prize for Sciences—you will find that the larger the bell the lower its pitch. If you make the size of the bell unwieldy, the pitch will become so low as to be inaudible. When the bell grows to the dimensions of the Earth, it will move only one swing in one hour, instead of more than a thousand swings in a second, which a house bell makes" (Pekeris 1958). The drum, the bell, and the Earth each have their typical equations of motion, and their defining boundary conditions, and a main task of the numerical approach is to approximate the values of the eigenfrequencies with their respective eigensolutions. This task fitted well into the overall horizon of interests of Pekeris and Alterman, particularly once they started to exploit the powerful capabilities afforded by WEIZAC.

Data gathered from studying free oscillations provided important information about the detailed structure of the Earth and about the density of its various internal layers, in ways that cannot be obtained from other seismic vibrations or waves. As such, from 1940 it became a fundamental part of seismology (Bullen 1975; Ben-Menahem 1995). Starting from scattered observations of earthquakes, the challenge in this area was to obtain a detailed specification of the distribution of the velocities of the waves within the Earth and of the disturbances at the source of the earthquake. But the theoretical roots of the relevant theories went somewhat earlier into the nineteenth century and the heyday of the theories of classical physics, and it is relevant to provide a brief account of some of the main ideas involved.

The origins of the study of the internal constitution of the Earth and the related phenomena of free oscillations are found in the theory of elastic vibrations of a solid sphere, going back at least to 1828, in the seminal work of Siméon Denis Poisson (1781–1840) (Poisson 1828). Fluctuations of an elastic ball were studied in the nineteenth century, when approximations were made for the oscillation cycle of the Earth. Important contributions came later on, mainly within the British tradition of fluid dynamics represented by William Thomson, Lord Kelvin (1824–1907), George H. Darwin (1845–1912) (Darwin 1879, 1882), Lord Rayleigh (1842–1919) (Thompson 1863), and also by Horace Lamb, already mentioned above (Lamb 1881).

Rayleigh developed a variational method meant to be used in investigating the behavior of waves as they propagate upon the free surface of an infinite homogeneous elastic solid. The disturbance undergone by these waves is confined to a superficial region of thickness comparable with the wavelength, which therefore behaves analogously to deep-water waves (Rayleigh 1885). Lamb, in turn, presented for the first time an elaborate mathematical model of an earthquake in a uniform semi-infinite elastic half-space arising from specified forms of initial disturbance.

The definitive exposition of Lamb's work appeared in 1892 in the classical *Treatise on the Mathematical Theory of Elasticity* of A. E. H. Love (1863–1940). It discussed in detail the simpler modes of vibration of a uniform sphere, classified the various general types of vibrations of a sphere, and calculated the roots of the so-called frequency equation (Love 1892). Somewhat later, in his classical work (Love 1911), he added his own important contributions to this field by incorporating into the theory the additional effects of gravity. Still, he relied on a highly simple and artificial model of the Earth in which the density, compressibility, and rigidity were all assumed to be constant. As will be mentioned below, such models would become more sophisticated in decades to come, and the results improve accordingly. Among other things, the slowest free oscillation period that he calculated was of nearly 60 min, a result that was considered to have little practical value given the difficulty of actually detecting ground movements with such long periods. Around 1950, however, the accuracy of the seismographic instruments was sensibly improved, to the extent that a period of about 57 min was actually detected. Below we return to the details of this important event.

The problem of oscillations is closely linked with that of the deformation of an elastic sphere, which can be caused either by surface forces or by internal body forces. A main feat of Love's book was that it subsumed all such cases under a similar analysis. Indeed, the problem of static deformation can be seen as the limiting case of an oscillation in which the period is infinitely long. Closely connected is the mechanical theory of small oscillations, which arise when a conservative system is slightly displaced from a configuration of "stable equilibrium." The mathematical models for these oscillations approximate in linear terms force laws that are inherently non-linear, and hence, their validity holds for a delimited range of states, which are close to equilibrium (Hestenes 1999, pp. 378–398). In the case of Earth oscillations, a complete account of the phenomena involved the superposition of a static deformation, a free oscillation (namely, an oscillation that arises with its own natural frequency and under no external influence other than the impulse that initiated the motion), and a forced oscillation. As the theories of the structure of the Earth in terms of layers continued to improve, the computational problems arising in the attempt to extend the results for a homogeneous sphere to an inhomogeneous model of the Earth proved to be more complex than initially expected. An initial breakthrough in this context came in 1926 when Robert Stoneley (1894–1976), working in Leeds, published an incomplete but suggestive draft of a possible variational technique to approach the calculations (Stoneley 1926). This approach would later be used and improved by AJP as the basis for their joint work with WEIZAC.

A main issue that was considered in all cases of static or dynamic deformations was the issue of scale. In "small bodies," i.e., those of the size of a minor planet, the effect of the mutual gravitational attraction of the bodies involved was considered to be minimal, whereas the elastic forces of cohesion in a solid were seen as predominant. To the contrary, the case of the Earth was taken as an instance of an elastic sphere for which the gravitational forces act as a significant correction to the elastic forces. Likewise, the oscillations of a fluid mass were considered to be of great importance in cosmological problems.

One particular problem that arose in studying a body of the size of the Earth from the perspective of the classical theory of elasticity was that, in small-scale problems, the deformations may always be regarded as small departures from an initially unstressed state. The situation can be expressed in terms of *stress* and *strain*. In general, the stress measures the force per unit area in a perfect fluid and the stress tensor has only one independent component, which corresponds to ordinary fluid pressure. The parameters of the tensor representing the strain, which is the deformation produced by the stress, correspond to the displacement associated with the deformation. When it comes to the interior of the Earth, the intensity of the state of stress that needs to be assumed, so that the ordinary theory of elasticity that may help calculating it, is incommensurably high and it requires introducing extra terms into the equations. Thus, the work done by the stresses during a small deformation gives rise to additional terms in the strain-energy tensor. Moreover, the state of initial stress must be known in order to get the full picture (Stoneley 1961).

In Lamb's theory, the sphere is treated as uniform and incompressible, and even for a sphere of the size of the Earth, the effects of gravitation and compressibility involve only a moderate correction to the results obtained for an incompressible solid. Thus, for example, the static deformation due to earth tides appears as a limiting case of the problem of harmonic oscillations as the frequency tends to zero.

If the Earth is seen as a body that is in a configuration of equilibrium, then the gravitational attraction can be seen as causing internal stress in the Earth. During oscillations, this body undergoes displacements from a configuration in which it is initially stressed. At any time, small additional stresses are superposed over the (presumably large) stresses that have been carried from the initial position and orientation of the configuration. The hypothesis is usually made that these "initial stresses" correspond to an initial state of hydrostatic equilibrium, not calculable by Poisson's theory of linear elasticity.

While most of the abovementioned considerations are related to general theories of the elastic sphere, there are also the specific conditions that relate to the internal constitution of the Earth—crust, mantle, inner and outer core—and specifically to the question of whether the inner core is solid or liquid. Whereas the classical nineteenth-century view assumed that the interior of the Earth comprised a thin crust floating on a molten interior, over the first decades of the twentieth-century seismological research had provided evidence of variation in the velocity of waves at different depths. Such variations were related to two elastic properties of a substance: (a) the *bulk* modulus, which measures the inherent resistance of the substance to compression, in terms of the ratio of the infinitesimal pressure increase to the resulting relative decrease of the volume, and (b) the *stiffness*, which measures resistance to deformation in response to an applied force, in terms of the ratio of the force applied and the displacement it produces in the substance. Seismological studies conducted just before WWI led to the insight that the Earth is made of solid rocks only down to a depth of 2900 km. All the rest constitutes what is known as "the liquid core of the Earth," with a radius of about of 3500 km.

More precise values of the density of the Earth's interior, which AJP will later rely upon in their 1959 article, were obtained on the basis of the joint work of Leason H. Adams (1887–1969) and Erskine D. Williamson (1886–1923) at the Geophysical Laboratory in Washington, DC. In a seminal article of 1923, they established the mathematical relationships between seismic wave velocities and the compressibility and density of rocks and asserted that "the dense interior cannot consist of ordinary rocks compressed to a small volume; we must therefore fall back on the only reasonable alternative, namely, the presence of a heavier material, presumably some metal, which, to judge from its abundance in the Earth's crust, in meteorites and in the Sun, is probably iron (Adams and Williamson 1923)."

Yet another important relevant contribution appeared in 1940–1942, in the work of Keith Edward Bullen, who developed a model consisting of seven concentric shells (Bullen 1940). In 1950, he presented a second model, Bullen's Earth-Model B, which incorporated results derived from a compressibility-pressure hypothesis (Bullen 1950). This model turned out to be highly useful for obtaining numerical values for the free periods of oscillation of the Earth. Bullen's ideas were further elaborated by the Cambridge geophysicist Edward Crisp Bullard (1907–1980), who, in an article of 1957, proposed six possible models (Bullard 1957). Pekeris and his collaborators would base their calculations on three of these models, referred to as Bullard model I, Bullard model II and Bullen model B. They did not differ greatly from one another, and, contrary to Bullen's model, in which the density of the inner core increases rapidly in the inward direction and has a heavier central core, Bullard's models increase only gradually (Bullen 1942, 1975; Lowrie 2007, pp. 93–197).

Increased interest in the study of free oscillations of the earth arose on the wake of a strong earthquake of November 4, 1952, that shook the peninsula of Kamchatka. The quake was recorded at the strain seismograph of the Caltech Seismological Lab in Pasadena. Established in 1926, the Seismo Lab, as it came to be known, became a world-leading institution from 1930 under the leadership of Beno Gutenberg (1889–1960) and Charles F. Richter (1900–1985), of Richter-scale fame.[5] Among its prominent members, the Lab also counted Hugo Benioff (1899–1968), who built his fame from very early on as an extremely ingenious and meticulous developer of precision tools.

Just one year prior to the Kamchatka quake, in November 1951, a journalistic article featuring the extraordinary capabilities recently developed at the Seismo Lab described the potential of that specific instrument, the strain seismograph, "one of the most valuable recording instruments" in a collection of several groundbreaking developments. The article indicated that:

> The response of this instrument is derived from strains produced in the ground by seismic waves rather than displacements of the ground, as is the case with all the pendulum-types of seismographs. The linear-strain seismograph records as little as six millionths of an inch of ground-squeezing produced by a distant quake, and if the Atlantic Coast should be squeezed a foot closer to the Pacific Coast it would record that. Observations made with this instrument, taken by themselves or in combination with those of the pendulum instruments, provide information concerning seismic waves which cannot possibly be had from the records of the pendulum instruments alone.[6]

Benioff had led the development of the strain seismograph and he was responsible for its operation (Benioff 1935). He discovered on the seismograms of the Kamchatka

---

[5] http://www.seismolab.caltech.edu/history.html.

[6] "Earthquakes Recorded on Tape," *Engineering and Science* 15 (Nov. 1951), 7–11.

earthquake of 1952 two long-period oscillations: one with a period of 57 min and another with a period of 100 min. Earthquake vibrations are typically measured at a rate of about 10–100 times per second, and occasionally also, oscillations of a period of about a minute are recorded. Benioff suggested that the newly recorded oscillations, with a period of the order of a whole hour, represent the "free spheroidal oscillation of the earth" (Benioff et al. 1953).

Benioff based his conjecture on a result appearing in Love's treatise of 1911, and already mentioned above. Love's calculations of the period of free spheroidal oscillation of the Earth were based, as already stated, on a homogeneous model of the planet. In addition, Love's analysis involved two specific assumptions about the so-called Lamé elastic coefficients, λ and μ, typically used to describe the behavior of a perfectly elastic material: (1) that the mean rigidity, $\overline{\mu}$, is equal to 8.9×10^{11} dyn/cm^2 and (2) that the values of the mean elastic coefficient $\overline{\lambda}$ equals that of $\overline{\mu}$. However, since the Earth was now assumed to be non-homogeneous, and the known values of the coefficients were $\overline{\mu} = 1.463 \times 10^{12}$, and $\overline{\lambda} = 2.402\overline{\mu}$. Benioff concluded that Love's originally calculated value of 60 min for the period was questionable. Benioff's measurements also helped lend a greater reliability to Bullen's model over that of Bullard.

Benioff's work led in subsequent years to the construction of seismographs of very long periods, specifically designed for identifying the kind of free oscillations that he had postulated (Press and Benioff 1958). At the theoretical side, as already stated, his work also led to an increased interest in the study of free oscillations in a more general sense (Benioff et al. 1954), and, in particular, it set the ground for the work of Pekeris and his associates. In fact, many had doubted the accuracy of the results reported by Benioff, but Pekeris and his group decided to try to calculate these fluctuations and to compare the peculiarities of Benioff's empirical data with the theoretical results obtained from calculating the vibration cycle for a realistic model of the Earth.

4.2 Pekeris and Geophysical Research in Israel

Over the years 1946–1947, still working at Columbia as director of the Division of War Research, Pekeris was involved in efforts to organize a Scientific Advisory Committee, comprising distinguished geophysicists and mathematicians, for the establishment of a seismological observatory in Palestine even before the creation of the state. Gutenberg was definitely enthusiastic about joining the committee. He stressed that research on seismic activity in the region, and in fact worldwide, would highly benefit from the construction of new observatories (a central one in Jerusalem and a number of secondary stations all around Palestine) that would join existing ones, such as that in Ksara, Syria, and Helwan, Egypt, as well as in Istanbul. Contrary to Pekeris's opinion, Gutenberg considered those centers to be "first-class or almost first-class."[7] And above all, Gutenberg had intrinsic, purely scientific reasons for supporting the promotion of earthquake research in this region of the world:

---

[7] Gutenberg to Pekeris, Apr. 12, 1946 (CPA).

There is no doubt that in the earlier history relatively large earthquakes have occurred there, and in all probability many small shocks would still be recorded by a sensitive instrument. As in most instances in regions with temporary decrease in seismicity, larger shocks may occur again in Palestine ... This observatory should also be a central place for applied seismology. The search for oil, valuable minerals and water by geophysical methods ... certainly would aid realty in the economic development. Dr. Pekeris, with his great theoretical experience, and a younger man with field experience, could easily bring the observatory to international importance.[8]

Pekeris, of course, stressed those same issues when writing to prospective members of the committee. Thus, in addition to "the recording and study of earthquakes, the observatory would also offer the Palestineans [sic] training courses in the theory and practice of geophysics which are required for geophysical prospecting."[9] This prospecting would be crucial for oil exploration in the country as well as for developing new water sources in regions of Palestine where supply was inadequate. These were surely important tropes in the Zionist ethos and both were high in Pekeris's own agenda of "applied geophysics" for Palestine.

Additional world-leading geophysicists who were eager to respond to Pekeris' invitation to join the committee included Maurice Ewing (1906–1974), from Columbia, Harlow Shapley (1885–1972), from Harvard, and Louis B. Slichter (1896–1978), from Wisconsin.[10] As usual, Pekeris went after the large picture and asked Weizmann to be personally involved in attracting additional scientists to enhance the status of the committee. Weizmann, for instance, asked Hyman Joseph Ettlinger (1889–1986), an American Jewish mathematician working at Austin, Texas, to help raise funds for the seismic observatory.[11] Pekeris also searched for additional funds at a meeting of geophysicists in Los Angeles, and from the Dutch Shell Corporation.[12]

Finding the right person to join Pekeris at Rehovot, however, proved to be a more challenging task. Pekeris offered a position at DAM to Arthur Erdélyi (1908–1977), then at Edinburgh, who expressed willingness to come to Palestine.[13] In the end, however, this offer did not materialize. Gutenberg and Benioff contacted some colleagues in the USA, but warned Pekeris that there "are few who are fit for this position, and the need for geophysicists in the United States is still very high." Moreover, while Benioff would be willing to advise in the construction of the observatory and pay sporadic visits to Palestine, he would not be willing to undertake a more direct involvement. Gutenberg also warned that there was a long backlog for obtaining the Benioff instruments, due to high demand, and Pekeris could not expect to get them any time soon.[14]

---

[8] Gutenberg to Ettlinger, Jan. 21, 1947 (CPA).

[9] Pekeris to Ettlinger, Dec. 13, 1947 (CPA).

[10] Ewing to Pekeris, Apr. 12, 1946; Shapley to Pekeris, Apr. 15, 1946; Slichter to Pekeris, Apr. 15, 1946 (CPA).

[11] Weizmann to Ettlinger, Jun. 26, 1946 (CPA).

[12] Pekeris invitation to Silverman, Sharp, Eissler, Mar. 10, 1947 (CPA).

[13] Pekeris-Arthur Erdélyi correspondence, Aug. 22, 1947 to Mar. 08, 1948 (CPA).

[14] Gutenberg to Pekeris, Dec. 18, 1946 (CPA).

Fig. 4.1 Frank Press (left) and others with the seismograph in Jerusalem. Photographer: Yehuda Eisenstark, courtesy of ISA

It was in late 1953, when, as part of an initiative of the Caltech Seismo Lab, the National Physical Laboratory (NPL) of the Research Council at the Prime Minister's Office agreed to install in Jerusalem and Safed two short-period seismographs equipped with photographic recording. Frank Press (1924–2020), then a young geophysicist, was called to supervise the works (Fig. 4.1).[15] Ari Ben-Menahem was trained to operate and maintain the stations. He started regular seismic recording, thus becoming the first instrumental seismologist in Israel. In 1958, Ben-Menahem left for graduate studies at Caltech and returned in 1965 as a professor of geophysics at WIS (Ben-Menahem 2003).[16]

Against this background, upon his arrival in Rehovot in 1949 Pekeris made sure that research activities in geophysics would start immediately. Much of this work was undertaken in cooperation with the government of the recently established State of Israel. The participants in a meeting held on August 1949 between the director of the Research Council of Israel, physicist Shmuel Sambursky (1900–1990), and representatives of WIS, including Pekeris and Israel Dostrovsky (1918–2010), head of WIS Isotopes Department, agreed upon cooperation in performing "the geophysical research needed to develop the country." The DAM established a geophysical laboratory aimed toward a geophysical survey of Israel, which, "incidentally," would help

---

[15] *Haboker* (Dec. 20, 1953).

[16] A seismic observatory was finally established in 1975, 12 km. North to the city of Eilat, thanks to the donation of the Brazilian publisher and philanthropist Adolpho Bloch.

in the search for water and oil.[17] A grant from the Ford Foundation also supported this project. The original proposal to the Ford Foundation included plans for a "Geological Survey, Water and Soil Conservation and Development of Water Resources."[18] Later, in the first comprehensive report submitted to the Foundation, two parts of the project were listed: a Gravity Survey and Application of Modern Automatic Computing Machinery to the Interpretation of Geophysical Measurements.[19]

Between 1951 and 1955, a major share of the activities of DAM were related to geophysical prospecting, through both the gravimetric method and the seismic method. In the seismic method, an "earthquake" is artificially produced and the echoes that bounce back from the various geological strata are recorded by seismographs. The depth of the interface between different kinds of rocks can be inferred from the time it takes for the waves to travel from the explosion, down to the boundary, and back. The gravimetric method is based on the assumption that if heavy rocks are buried underground, then their presence can be detected on the surface by the extra gravitational pull it exerts on a pendulum. This gravitational effect, moreover, can be measured with the help of very sensitive instruments. Over oilfield layers, the rocks at depth are not horizontally flat, but rather bent upward in the shape of a dome. Thus, the heavier deep rocks are closer to the surface in the top of the dome, and at those points, the measurements will indicate the presence of a stronger gravitational pull (Pekeris 1964, p. 3).

The geological survey started in the summer of 1950. The prospects of finding oil in the Promised Land attracted much public attention, and it was widely published between 1951 and 1953 in the daily press in Israel as well as in the USA.[20] Even the Yiddish press announced with joy "Oil in Israel: from the present difficulties to a great future."[21] The weekly "Dvar Hashavua," in its issue of May 22, 1952, for example, published a lengthy interview with Pekeris and explained the basics of the survey in the following words:

> The survey starts with the gravimetric instrument, which is able to measure the gravitational effect of the various soil layers. When it comes across "suspicious" layers, then the second survey is applied, following the seismic method, which is able to find out deep in the earth those layers possibly containing oil. ... The seismic department works with a drilling machine, able to dig in two hours to a profundity of 60–100 m. This department makes up to three drills a day. The drills are separated 250 m from each other. When the drill is completed, 20 Kg, of seismic TNT are pushed into the bottom to a height of three meters. The rest of the drill is filled-up with water. Twenty-four seismometers are placed around the drills, 12 on each side, and they measure the in-ground shock. The seismometers are connected to a cable, which in turn is connected to a seismic truck, that records the ground oscillations at high-speed, and drafts a diagram of the oscillations.[22]

---

[17] Scientific Activity Report, 1949 (WIA).

[18] Proposal to Ford Foundation, Aug. 21, 1952 (ISA 4411 10/393).

[19] Israel Foundation Trustees, First Comprehensive Report, Jun., 1955 (ISA 4381 2520 ‎א/1).

[20] See, e.g., *Haaretz* 25–09-1951, p. 4; *B'nai Birth Messenger* (L.A. California) Sept. 11 9153, p. 1; *New York Times June* 6, 1952, p. 6.

[21] A newspaper cut-out found in Pekeris' archive at WIS (CPA) bears the title: "‏אויל אין ישראל: פֿון די‏" ‏איצטיגע שוועריגקייטן צו א גרויסם צוקונפט‏". Unfortunately, we have not been able to identify the original source, reference details, and date of publication.

[22] Translated from Hebrew by the authors.

A leading figure in the geological survey was the mathematician Joseph Gillis (1911–1993), who joined DAM in 1948 after working as cryptographer at Bletchley Park during WWII. He led the gravimetric project, and eventually, Avihu Ginzburg (1926–2017) took over from him. Ginzburg was to complete his Ph.D. in geophysics in 1961 under Pekeris's guidance and later became one of the first professors in the Department of Geophysics at Tel Aviv University (which became the only department in Israel with a complete educational program in geophysics). The leading figures in the seismic part were Shaul Meiboom (who became a pioneer in research on nuclear magnetic resonance, see below) and Michael Behr (who went on to work at the Burroughs Corporation, California). The survey covered an area of close to 8000 km^2.

And then, on September 23, 1955, oil was finally found in the Heletz field at a depth of 1515 m. Heletz was a Jewish settlement established in 1950 near Ashkelon, where the Palestinian village of Huleikat was previously located. In 1933, the zone had already been explored as part of a preliminary survey initiated by the British Mandate authorities and that was abandoned in 1948 with the outbreak of war. Now, the discovery immediately became an event of national importance and international resonance. British Foreign Office officials hastened to define it as "the most important event that has happened in Palestine since [Independence Declaration Day] May 15, 1948" (Bialer 1998, 190–91). Pekeris sent telegrams to both von Neumann and Oppenheimer: "Oil discovered today in Israel."[23] Von Neumann immediately answered: "heartfelt congratulations to great success to which your work and guidance has so decisively contributed."[24]

The local press festively reported: "The discovery of oil in Israel arouses resonant echoes throughout the world—reported the daily "Davar". Oil found in Heletz is of high quality and there are considerable quantities of it."[25] A Hebrew song was released on the Israeli radio and an LP record that soon became very popular. It called on women and men in the country, in candid Zionist pathos typical of the time, to go out dancing, from Dan in the North to Eilat in the South, and to sign in full voice: "Oil is flowing at Huleikat! Oil! Oil! Oil! Oil! Oil is flowing at Huleikat!"[26]

Pekeris was highly optimistic that oil would be extracted in commercial quantities throughout the country (Fig. 4.2), starting from Heletz and then north up to Haifa.[27] He later suggested that Israel was the only country in the world where a complete gravitational survey had been carried out with the precision required for oil prospecting (Pekeris 1958). All expectations, however, turned out to be exaggerated and the Heletz oil field came to provide only a tiny percentage of the country's overall consumption and for a rather short time at that. One way or another, in July 1957 it was decided that the teams working on the ground had become experienced

---

[23] Pekeris to von Neumann, Sep. 23, 1955 (CPA).

[24] Von Neumann to Pekeris, Sep. 25, 1955 (CPA).

[25] *Davar* (Sep. 25, 1955).

[26] "Oil is flowing in Huleikat!" (נפט זורם בחוליקת); https://youtu.be/oCFpa30leQs. Lyrics: Haim Shalmoni; Music: Shlomo Weisfish.

[27] *Haboker* (Oct. 1955).

Fig. 4.2 Pekeris studying charts of echo soundings to determine the best spot for drilling operations at "Heletz". Courtesy of the Israel National Photo Collection (*Credit* Pridan Moshe)

enough to be able to continue on their own. A state-run company was established, the Geophysical Institute of Israel, and Pekeris and his collaborators continued to work with them as advisors.

WEIZAC became operational in 1955. Among the earliest tasks assigned to it was the computation of data related to a "residual gravity map." That same year, Alterman joined the DAM at WIS after receiving her Ph.D. from the Hebrew University in Jerusalem. Alongside Alterman, Hans Jarosch also collaborated with Pekeris in his geophysical research. He joined the WEIZAC team in 1954, right after finishing military service. A mechanical and aeronautical engineer by training, he had gained experience as a mathematical assistant in the EDSAC project. He had arrived in the UK in 1938, aged 10, as part of the famous *Kindertransport* rescue effort. In 1951, he immigrated to Israel and joined the Air Force. Pekeris employed him initially as a "human computer" working with a Marchand machine as well as assisting Zvi Riesel (1922–2002), another central figure of the WEIZAC project (Corry and Leviathan 2019, pp. 51–52), in the task of writing check-up routines. After the publication of the AJP joint work, Jarosch went on to complete a Ph.D. degree at Queen Mary and Westfield College, University of London, with a dissertation that dealt with the oscillations of the earth.[28]

This was a time, then, when various circumstances converged and stood at the background of the AJP collaboration. Alterman's detailed reports to Pekeris show that WEIZAC was being used at the time not only for research on the Earth's oscillations, but also for work on "Air Tides," and "Ocean Tides," collisions of elastic solids, boundary value problems for ODEs and PDEs, and the writing of general-purpose routines that were needed in various contexts.[29] The availability of WEIZAC and the gradual mastering of its use at DAM took place at a time when groundbreaking

---

[28] Jarosch, interview by Leviathan, Nov. 11, 2010.

[29] Alterman, "Summary of work done during December 1957," in "Progress Reports for the Years 1956–57," pp. 115–119 (CPA).

technologies appeared not only in the field of electronic computing, but also, among others, in the field of highly sensitive seismic instrumentation. Large amounts of seismological data were gathered from all over the world, and at the same time, important theoretical advances were achieved. Pekeris could work with two talented and highly motivated, younger colleagues. In 1955, his own interest led him to publish two articles on analytic solutions of equations related to geophysical questions (cited above). The time was ripe for devoting energies to the study of free oscillations. Pekeris and his colleagues at WIS developed a canonical formulation of the problem, which turned out to be ideal for numerical calculations, and the availability of WEIZAC created a unique situation for a breakthrough.

4.3 Alterman, Jarosch, and Pekeris on Earth Oscillations

The AJP 1959 article took as its starting point a concise synthesis of both theoretical knowledge and empirical data that recent seismographic and geophysical research had made available, and that was briefly summarized in the foregoing pages. On the occasion of being awarded the Weizmann Prize in 1958, Pekeris summarized the aims of his geophysical research with the following words:

> The purpose of our investigations, which were carried out on the electronic computer of the Weizmann Institute, was first to test the validity of Benioff's hypothesis, and secondly to see whether the observed periods are sensitive to the internal constitution of the Earth. The first result which Mr Jarosch and I obtained was that if we assume the Earth to be solid from the surface down to the center, the period of free oscillation turns out to be only three quarters of an hour, rather than the observed value of close to a whole hour. This result lends additional support to the now accepted view of the existence of a liquid core in the Earth. Subsequent investigations, in which Dr Z. Alterman joined us, showed that if we do assume the existence of a liquid core, then both Bullen's and Bullard's models give a period of free oscillation of about 54 min, which agrees within the experimental error with the observed value of 57 min. The 57 min oscillation therefore provides no criterion for choosing between the two models. (Pekeris 1958, p. 258)

The aim of AJP's calculations was to determine not only the period of spheroidal oscillations, but the whole spectrum of the free oscillations of the Earth, comprising also the radial and torsional ones. The main question they asked was whether, and to what extent, the free period, especially of the spheroidal oscillation, is sensitive, as Pekeris explained, to variations in the internal constitution of the earth.

AJP made it perfectly clear that the complex calculations required to achieve these aims were made possible only by the availability of WEIZAC. They pursued their calculations in relation to various models of the constitution of the earth: (1) Love's homogeneous model; (2) a homogeneous solid mantle enclosing a homogeneous liquid core; (3) Bullen's model B, and (4) Bullard's models I and II. They took the properties of models (1) and (2) to agree with the average values of the real Earth in the respective regions.

For all models considered, except for the homogeneous one, the calculations yield spheroidal oscillations of order $n = 2$, with a period of about 53.5 min. For the homogeneous model, they obtained a period of only 44.3 min. Thus, Benioff's measurement for the Kamchatka earthquake, with a period of 57 min agreed well, up to observational error, with the calculation but was not sufficiently sensitive to distinguish and adjudicate among the various models (except, again, for the homogeneous model, which was rendered inacceptable). Benioff's second long-period oscillation, for which he measured a value of 100 min, also agreed with AJP's calculation in the case of Bullen's model B, which in addition displayed a new kind of spheroidal oscillation of order 2, the "core oscillations," with a period of 101 min. Such oscillations did not arise in the calculations related to the other models, but it was confined mainly to the core, whereas in the mantle its amplitude became negligible.

Benioff's measurements might thus be seen as possible evidence in favor of Bullen's model B. Still, a significant difficulty arose in this case, which weighed against accepting the calculations as evidence in favor of Bullen's model B over Bullard's two models. The difficulty relates to the impossibility to excite a free vibration by applying a force at a nodal point. Stoneley had shown in 1931 that the deeper the focus of an earthquake the smaller the amplitudes of the surface (free) waves (Stoneley 1931). Since in the case of core oscillations the whole mantle is a nodal region, it would be difficult to understand how the Kamchatka earthquake could have excited it. So, one of the points addressed in the article concerned the determination of the relative amplitudes of the various free modes of oscillation, which would be excited by a compressional point source situated at a shallow focal depth. And the results indeed indicated that the amplitude of the core oscillation should be only of the order of 0.1% of the amplitude of normal oscillations.

In the following two subsections, we present a somewhat more technically detailed description of the way in which AJP formulated the problem in mathematical terms that could be used to calculate the desired values with the help of WEIZAC, and how the RK methods were put into practice in this specific context.

4.3.1 AJP's Formulation of the Oscillations Problem

In order to formulate the problem in a manner more amenable to computations made on WEIZAC, AJP had to further develop the theoretical considerations on which their study was based. The main magnitudes related with these considerations are V_p, V_s, denoting, respectively, the velocities of the compressional (or primary) and shear (or secondary) waves, both of which are functions of the radius r. These magnitudes are measured empirically, mainly from earthquake data, and their values are determined by properties of the earth such as the density ρ and the adiabatic bulk modulus k, both of which are also functions of the radius r. In addition to their empirical basis, V_p, V_s are mathematically related by the condition:

$$\frac{k}{\rho} = V_p^2 - V_s^2.$$

Relying on the work of Adams and Williamson in 1923 (mentioned above), and assuming with them that the internal pressure of the Earth is essentially hydrostatic and its variation is due only to compression and not to change of material, the density $\rho(r)$ can be determined from the following two equations:

$$\frac{d\rho}{dr} = -\frac{Gm\rho}{\pi^2\left(V_p^2 - V_s^2\right)},$$

$$\frac{dm}{dr} = 4\pi r^2 \rho.$$

The additional parameter $m(r)$ denotes here the mass comprised within the sphere of radius r, and the constant G is, as usual, the universal gravitational constant.

These two equations, together with empirical data provided by previous measurements of the seismic velocity distributions, including those of the Caltech Lab, provided the basis for the models proposed by Bullen and Bullard. AJP plotted their results in a table that is reproduced here (Fig. 4.3). Of the six models considered by Bullard, they chose models I and II as being the two extremes. Bullen's model B differs from the models of Bullard, as already indicated, by having a heavier central core, as reflected in the graph.

Now, while the seismic data was taken to yield accurate information on the detailed distribution of *velocities* inside the earth, the distribution of *density* was only indirectly known, and hence, it could be seen as a significant source of uncertainty. This was a main motivation for looking for new, independent, and much more accurate

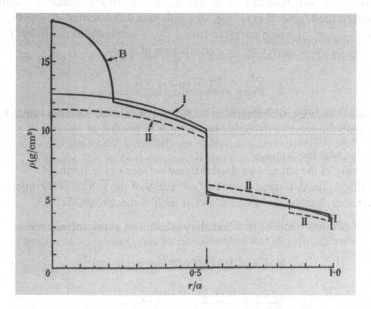

Fig. 4.3 Density distributions of Bullen's model B and of Bullard's models I and II, as plotted in (Alterman et al. Oscillations of the Earth, 1959, p. 85). The arrow ↓ indicates the core boundary. Used with permission of The Royal Society (U.K.); permission conveyed through Copyright Clearance Center

observational data, and this was the point where Benioff's recent measurements came to play a central role. Benioff's observations were assumed to be correct concerning the distribution of density, but with the proviso that, until they were further corroborated by new observations, they should be taken with caution.

AJP started by considering three basic rock properties $\rho_0(r)$, $\mu(r)$, $\lambda(r)$. The first one, $\rho_0(r)$, denotes the unperturbed density at distance r, while the other two, μ, λ, are the already mentioned Lamé coefficients. Following Love's classical account, these properties are related via the following two equations:

$$V_p^2 = \frac{\lambda + 2\mu}{\rho_0}; \quad V_s^2 = \frac{\mu}{\rho_0}.$$

There is also a displacement function, denoted here as a vector **u**, which is expressed in spherical coordinates

$$u_r = u, \quad u_\theta = v, \quad u_\phi = w.$$

In these terms, three Laplace-like equations of motion can be formulated for

$$\rho_0 \frac{\partial^2 u}{\partial t^2}, \quad \rho_0 \frac{\partial^2 v}{\partial t^2}, \quad \rho_0 \frac{\partial^2 w}{\partial t^2},$$

each of which is equated to a somewhat complicated, linear differential expression (which we skip here) involving four additional magnitudes, together with their first-order derivatives with respect to r, θ, and ϕ. These are the dilatation coefficient Δ, the unperturbed force of gravity g_0, the gravitational potential ψ, and the strain components e_{ij}, all of which are taken from Love's treatment of elasticity. There is also a fourth equation governing the perturbation of ψ,

$$\nabla^2 \psi = 4\pi G(\rho_0 \Delta + u \dot{\rho}_0),$$

with the dot indicating differentiation with respect to r. In addition, three further conditions are imposed on the possible solutions of these four equations:

(1) Regularity at the origin,
(2) Vanishing of the stresses on the deformed surface of the Earth,
(3) Equality at the deformed surface of the Earth of the values of the internal and external gravitational potentials, and of their respective gradients.

These conditions translate into boundary conditions at the surface $r = a$:

$$e_{r\theta} = e_{r\phi} = 0, \tag{4.1}$$

$$\frac{\partial \psi_n}{\partial r} + \frac{(n+1)}{a} \psi_n = 4\pi G \rho_0 u_n. \tag{4.2}$$

Here, the subscript n denotes the spherical harmonic component of order n that naturally arise in the solution of Poisson equations on spherical surfaces. An additional important point is that at the core boundary—which should be seen as an internal surface of discontinuity—the continuity of the stress components u and ψ must be postulated, and this is done in terms of the following equation:

$$\frac{\partial \psi_1}{\partial r} - \frac{\partial \psi_2}{\partial r} = 4\pi G(\rho_{01} - \rho_{02})u, \tag{4.3}$$

where the indexes, 1, 2, denote the two sides of the interface.

4.3.2 Oscillations and Runge–Kutta with WEIZAC

The first kind of oscillations calculated by AJP was *purely torsional oscillations*. Such oscillations arise when both the dilatation Δ and the radial component u_r of the displacement vector $\mathbf{u}$ vanish. If, in addition, the density remains constant and the radial displacement of the boundaries is null, then no perturbation arises in the gravitational field, and motion is therefore controlled entirely by elastic forces. Moreover, because of the liquidity of the core, this motion is confined to the rigid mantle.

In this case, the somewhat complicated Laplace-like equations of motion led to a relatively simple solution, which AJP assumed to be of the following form:

$$u = 0, \quad v = \frac{V(r)}{\sin\theta} \frac{\partial S_n(\theta, \phi)}{\partial \phi} e^{i\sigma t}, \quad w = -V(r) \frac{\partial S_n(\theta, \phi)}{\partial \theta} e^{i\sigma t}, \tag{4.4}$$

where V is the tangential displacement and $S_n(\theta, \phi)$ denotes a spherical harmonic of order n. This solution satisfies identically the component $\rho_0 \frac{\partial^2 u}{\partial t^2}$ in the equation of motion, while for the components $\rho_0 \frac{\partial^2 v}{\partial t^2}$, $\rho_0 \frac{\partial^2 w}{\partial t^2}$, an additional condition is required, involving the radial displacement U, namely:

$$\mu\left(\frac{d^2 U}{dr^2} + \frac{2}{r}\frac{dU}{dr}\right) + \frac{d\mu}{dr}\left(\frac{dU}{dr} - \frac{U}{r}\right) + \left[\sigma^2 \rho_0 - \frac{n(n+1)\mu}{r^2}\right]U = 0. \tag{4.5}$$

Now, recall that variable $\mu(r)$ is empirically determined. This creates an awkward situation in terms of the numerical calculation of the solution of Eq. (4.5), which involves its derivative. Thus, AJP introduced two independent variables that helped bypass this difficulty, namely

$$y_1 = U; \quad y_2 = \mu\left(\frac{dU}{dr} - \frac{U}{r}\right), \tag{4.6}$$

where y_2 is the radial component of the shear stresses $\tau_{r\phi}$, $\tau_{r\theta}$. Stated in this terms, Eq. (4.5) turns into the differential system:

The iteration converged to eight places at least, but only about three places are significant as the data in Bullen's tables are given only to three places. The final results are, for the nth harmonic:

	n = 2	n = 3	n = 4
Fundamental	44 min 6 sec	28 min 33 sec	21 min 52 sec
I overtone	12 min 42 sec	11 min 36 sec	1o min 29 sec
II overtone	7 min 2o sec	7 min 7 sec	6 min 52 sec

Fig. 4.4 Alterman's report to Pekeris on results obtained with WEIZAC for the equations of the Earth's free oscillations, December 1957 (CPA)

$$\frac{dy_1}{dr} = \frac{1}{r}y_1 + \frac{1}{\mu}y_2, \tag{4.7}$$

$$\frac{dy_2}{dr} = \left[\frac{\mu(n^2 + n - 2)}{r^2} - \sigma^2\rho_0\right]y_1 - \frac{3}{r}y_2. \tag{4.8}$$

Here the boundary conditions are that at $r = a$ and at $r = b$ (the radius of the core) we have $y_2 = 0$. And now, Eqs. (4.7) and (4.8) lend themselves to integration by Runge–Kutta methods, and this is where AJP used WEIZAC to determine the periods of free torsional oscillations of spherical harmonic orders $n = 2, 3$ and 4 for Bullen's model B, as well as the mass-average properties arising in this model. In Fig. 4.4, we reproduce the results that Alterman obtained in December 1957 and communicated to Pekeris in her periodical reports.[30]

The second kind of free oscillations discussed in AJP's article is the *spheroidal oscillations*. They arise when the radial component of the curl of the displacement, u_r, vanishes, while the corresponding additional components u_θ and u_ϕ, are different from zero. In this case, the equivalents of the three expressions in Eq. (4.5) are these:

$$u = U(r)S_n(\theta, \phi), \, v = V(r)\frac{\partial S_n(\theta, \phi)}{\partial \theta}, \, w = \frac{V(r)}{\sin\theta}\frac{\partial S_n(\theta, \phi)}{\partial \phi}, \tag{4.9}$$

each of them multiplied by a common time factor, $e^{i\sigma t}$. Given that the dilatation Δ does not vanish in this case, gravitational forces do come into play, and the gravitational potential is affected by the gravitational perturbation Ψ, which is expressed as follows:

$$\Delta = X(r)S_n(\theta, \phi), \quad \Psi = P(r)S_n(\theta, \phi), \tag{4.10}$$

---

[30] Alterman, "Summary of work done during December 1957," in "Progress Reports for the Years 1956–57," p. 4 (CPA).

where P is the amplitude of the perturbed gravitational potential, and

$$X = U + \frac{2}{r}U - \frac{n(n+1)}{r}V.$$

Like in the case of the purely torsional equations, the three expressions appearing in (4.9) are now substituted in the three components of the equations of motion for $\rho_0 \frac{\partial^2 u}{\partial t^2}, \rho_0 \frac{\partial^2 v}{\partial t^2}, \rho_0 \frac{\partial^2 w}{\partial t^2}$, yielding three rather complex expressions, which represent three simultaneous second-order differential equations for the three functions U, V, P, and that need to be solved subject to the boundary conditions represented by (4.1–4.3).

Once again, it is necessary to bypass the difficulty created by the fact that the second-order equations involve magnitudes that are empirically determined, in this case ρ_0, μ, and λ (respectively: the unperturbed density at distance r, and the two Lamé coefficients). In order to do so, AJP introduced the following six dependent variables:

Once again, it is necessary to bypass the difficulty created by the fact that the second-order equations involve magnitudes that are empirically determined, in this case ρ_0, μ, and λ (respectively: the unperturbed density at distance r, and the two Lamé coefficients). In order to do so, AJP introduced the following six dependent variables:

$$y_1 = U, \quad y_2 = \lambda X + 2\mu \dot{U}, \quad y_3 = V,$$

$$y_4 = \mu\left(\dot{V} - \frac{V}{r} + \frac{V}{r}\right), \quad y_5 = P, \quad y_6 = \dot{P} - 4\pi G \rho_0 U.$$

In this way, the three second-order differential equations are transformed into six simultaneous linear equations on the six new variables for the mantle. In the liquid core, the situation is reduced to solving five simultaneous linear equations, since there the following supplementary conditions hold:

$$\mu = 0, \quad y_2 = \lambda X, \quad y_4 = 0.$$

After indicating some additional conditions that hold at the surface and at the boundary between the core and the mantle, AJP could solve the eleven equations with WEIZAC, using Runge–Kutta methods. Some numerical results and diagrams representing solutions are reproduced in Table 4.1 and Fig. 4.5.

Here, model α consists of a homogeneous solid mantle enclosing a homogeneous liquid core, with properties equal to the averages in the respective regions of Bullen's model B. Model β, in turn, is meant to represent a homogeneous solid with properties equal to the average over the whole globe of Bullen's model. The overall picture that arises from the calculations is that the fundamental period for $n = 2$ is close to 53.5 min. For model α it is within 25 min of this figure. For the homogeneous model β, the period is only 44.3 min, too far removed from Benioff's observed value of 57 min. Thus, with the exception of this latter model, the periods for the other four models are rather close in all the cases computed. AJP indicated that similar results

Table 4.1 Values for the periods of the spheroidal oscillations (in minutes) for the various models

n		Bullen B	Bullard I	Bullard II	α		β
0	Fundamental	20.0	–	–	–	–	26.7
	Overtone I	10.0	–	–	–	–	10.4
2	Fundamental	53.7	53.4	53.2	53.2	56.0	44.3
	Overtone I	24.7	24.8	24.0	24.0	25.2	–
	Overtone II	15.5	15.5	15.0	15.0	16.3	–
	Overtone III	9.8	9.8	–	–	10.4	–
	Overtone IV	8.0	8.0	–	–	7.9	–
3	Fundamental	35.5	–	35.0	35.0	–	–
	Overtone I	17.9	–	17.4	17.4	–	–
	Overtone II	13.6	–	13.1	13.1	–	–
4	Fundamental	25.7	–	25.3	25.3	–	–
	Overtone I	14.4	–	14.0	14.0	–	–
	Overtone II	–	–	11.8	11.8	–	–

Reproduced from (Alterman et al. 1959, p. 88), with permission of The Royal Society (U.K.); conveyed through Copyright Clearance Center

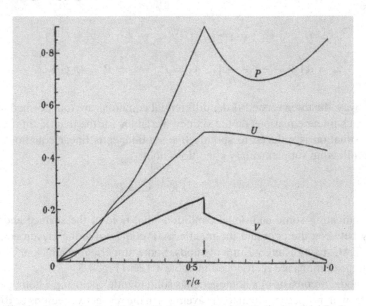

Fig. 4.5 The fundamental spheroidal oscillations for $n = 2$, in Bullen's model B: P = amplitude of the perturbed gravitational potential, U = radial displacement, V = tangential displacement. The arrow ↓ indicates the core boundary. Reproduced from (Alterman et al. 1959, p. 88), with permission of The Royal Society (U.K.); conveyed through Copyright Clearance Center

for the fundamental mode of $n = 2$ had been recently obtained, using a variational method and applying Rayleigh's method, by Jobert (1957).

Figure 4.5 shows, in turn, the distribution of the perturbed gravitational potential P, the radial displacement U, and the tangential displacement V for the fundamental oscillation in the case $n = 2$ in Bullen's model B. At the core boundary, U is continuous, P is also continuous (though its gradient is discontinuous there on account of the discontinuity in density), while V is discontinuous.

4.4 Aftermath

The numerical solutions obtained with WEIZAC allowed AJP to analyze the details of the data obtained in empirical measurements by Benioff and the extent of their correspondence with the various existing theoretical models, particularly in relation to the periods of free oscillation obtained in each model. Of particular importance was the "core oscillation," namely the free oscillation found for a period of 101 min. They admitted that identifying this core oscillation with the oscillation of a period of 100 min observed by Benioff in the Kamchatka earthquake of 1952 was indeed suggestive, but it still met with certain theoretical difficulties concerning its excitation. As a matter of fact, for a shock generated by a compressional point source situated at a shallow focal depth the amplitude of the core oscillation would be only of the order of 0.1% of the amplitude of normal oscillations. This indicated that the coincidence of periods might probably be accidental, and a number of free oscillations with periods ranging between one hour and several minutes could be excited with comparable amplitudes. Thus, they suggested, a task that remained open was that of searching for these long-period oscillations in earthquake records.

In an interim report of 1956, Pekeris and Jarosch had also declared that:

> Pending the substantiation of this result by an exact solution of the problem currently in progress on the WEIZAC, it would seem provisionally that the existence of the liquid core as well as the effect of the constraints imposed by the boundary conditions at the surface, where the rigidity is smaller than the average, tend to make the real earth behave dynamically in spheroidal vibrations as a less rigid body than is represented by the mass-average value of its rigidity. (Pekeris and Jarosch 1956, p. 24)

To get an exact result using WEIZAC, the final answer should be at least as accurate as the accuracy of the input data. Moreover, in order to bypass considerations of scaling and for ease of programming, it was decided to use the interpretive system with Double-Precision Floating (Alterman and Rabinowitz 1958).

Finally, summarizing their article, AJP stressed that the importance of their contribution went well beyond the specific conclusions and results presented there. The more general achievement, they indicated, was that by having already programmed the electronic computer for this task, the treatment of any new model that might be proposed would involve merely entering the new data into the program. Thus, as Jarosch indicated in his thesis, the program originally written for the computation of the periods of free oscillations could be easily adapted to also solve the problem of Earth tides and that of excitation of free oscillations by a point source (Jarosch 1962).

When the AJP article was published in 1959, the authors warned that the data provided by Benioff's work, which served as the basis for comparing theory with calculations, must be taken with caution until further corroborated by new observations. The Great Chilean Earthquake of 1960 became a source of empirical data of a quality and accuracy that was theretofore unknown. The 12th International Conference of the Geodesy and Geophysics was held three months after the earthquake. The collaboration between the WIS group and the Caltech Seismo Lab was insistently stressed (Benioff et al. 1961), and, at the same time, this important event turned out to be rather dramatic as various research groups reported their measurements of the post-earthquake fluctuations (Helsinki 1960, pp. 469–470). Retrospectively, AJP described the background to the meeting in the following terms:

> When Benioff announced in 1954 that he had observed an oscillation of 57 min on the seismogram of the Kamchatka earthquake of 1952 and suggested that it might be the free oscillation of the earth, there was no valid objection that could be leveled against those theoreticians who chose to adopt a wait-and-see attitude. This isolated observation, made for the first time after a century of seismometry, did not seem a sufficiently compelling reason for undertaking the cumbersome task of solving the system of coupled elastic and gravitational equations which govern the free oscillations of the earth, especially if the latter is to be represented by a realistic model based on empirical geophysical data. ... The final formulation of the problem of spheroidal oscillations was given in terms of a system of six simultaneous first-order differential equations. This system is particularly suited for integration on an electronic computer by the Runge–Kutta method, and it was chosen so as to avoid the need for differentiating explicitly the empirically determined elastic constants $\lambda(r)$ and $\mu(r)$, and the density $\rho(r)$. The period of free oscillation in the basic spheroidal $n =$ 2 mode came out around 53.5 min for several models, agreeing within the experimental error with Benioff's observed value of 57 min. ... As the years passed and no further registrations of the earth's free oscillations were reported, doubts began to arise as to the reality of the original observation in the Kamchatka earthquake of 1952, especially since the experimental techniques were continually improving. (Pekeris et al. 1961, p. 91)

But the time was now ripe for more conclusive results. First came an announcement by Frank Press, then at the Seismo Lab (and later president of the National Academy of Sciences and science advisor to President Carter), that together with Benioff they had recorded not merely a single long-period ground motion, but a whole series caused by the earthquake. Press claimed that the records were of free Earth oscillations, much as Benioff had claimed back in 1952, following the Kamchatka earthquake (Benioff et al. 1961). Three other groups then announced that they had likewise recorded similar series of periods, the longest period in all cases being about 54 min. After some deliberations meant to compare the details of the respective measurements, it turned out that there was a very good agreement among them. One exception was in the number of periods, which was lacking in the work of Slichter's group. These measurements had been performed on a tidal gravimeter of the "Lacoste-Romberg" type, a tool not primarily intended for recording earthquakes. Pekeris explained the differences that had arisen by indicating that the gravimeter would record only a specific kind of oscillations ("the spheroidal and not

Fig. 4.6 A first-day service envelope sent to Leah Pekeris by the Israel Postal Service, celebrating "Israel's contribution to science"

the toroidal"), whereas the seismographs of Press and Benioff would measure both. Pekeris's calculations showed that the results of the theoretical model complied with the measurements, hence confirming that all the groups had indeed recorded genuine natural long-period oscillations of the whole Earth (Brush 1996, pp. 199–200).

A report of this important episode in the *New York Times* emphasized the importance of Pekeris's contribution:

SCIENTISTS DETECT EARTH VIBRATIONS; First Such Observation Is Noted in 4 Laboratories After Chilean Quakes Resonance Is Forecast - Israeli Professor Furnished Calculations …The model of the earth was constructed by Dr. Keith E. Bullen of the University of Sydney Australia, one of the world's foremost authorities on the earth interior, who is in attendance here. The calculation of its resonant vibrations was made by Dr. Pekeris of the Weizmann Institute of Science in Israel.[31]

Following this success, the subject of free oscillations of the Earth became a major branch of observational seismology. The AJP article was repeatedly cited and Pekeris became a truly major figure in the field.[32] Pekeris coined the term "Terrestrial Spectroscopy," which became standard (Freeman 2004). On Independence Day of 1964, the State of Israel issued a stamp "Israel's contribution to science: Terrestrial Spectroscopy" (Fig. 4.6).[33] Alterman joined Tel Aviv University in 1967, where she established the Department of Environmental Sciences (later the Department of Geophysics and Planetary Sciences) and was head of this department almost until the day of her premature death in 1974.

---

[31] Salivan, W., "Scientists detect earth vibration" (*New York Times*, Aug. 8, 1960).

[32] Google Scholar indicates more than 600 citations. For details on the immediate impact of the article, see (Slichter 1967). For later developments, see (Woodhouse and Deuss 2005).

[33] Three stamps were issued for the 16th independence day of Israel titled: macromolecules of the living cell, Terrestrial spectroscopy, and electronic computer. (israelphilately.org.il/he/catalog/stamps/706/יום העצמאות הששה עשר).

4.5   Ocean Tides

In 1960, on the occasion of the General Assembly of the International Union of
Geodesy and Geophysics held in Helsinki (where the AJP article on free oscillations stood at the center of interest, as discussed above in Sect. 4.4), Pekeris
presented an NSF-funded, joint work with Menachem Dishon (who would later
become MAMRAM's commander), which comprised the first global numerical
solution of the Laplace tidal equations, involving realistic shorelines and bottom
topography. Four years after having "near succumbed to the general feeling that this
task was impossible," they were able to come up with accurate results, successfully
calculated with WEIZAC. The tides calculated for the Red Sea and the Black Sea
were in full accordance with recent observations as well as with existing analytic
solutions known for relatively simple cases. Their announcement was received with
great enthusiasm by the community of oceanographers, who expected that the new
approach would soon be applicable to the entire oceanic system (Saint-Guily 1961,
p. 370), though the full details were not disclosed at that time.

Research on tides and the questions related to the oscillations of the Earth had
originated within one and the same tradition described above (Sect. 4.1), involving
the likes of Lord Kelvin, Lord Rayleigh, and Lamb. As indicated in Sect. 4.3.1,
Jarosch had stressed in his Ph.D. thesis of 1962 that the computer program originally
written for computing the periods of free oscillations with WEIZAC could be easily
adapted to solve problems related to tides. And indeed, already by 1964 Pekeris
and Accad had published some preliminary results, partly based on calculations with
WEIZAC and then completed with the CDC-1604 (Accad and Pekeris 1964). For the
1969 joint article with Accad (Pekeris and Accad 1969), as well as for a second joint
paper (Accad and Pekeris 1978)—two publications that continued to be frequently
cited for decades to come—calculations were conducted on the GOLEM.

References

Accad, Y., and C.L. Pekeris. 1964. The K2 tide in oceans bounded by meridians and parallels.
    Proceedings of the Royal Society of London. Series A, Mathematical and Physical Sciences 278
    (1372): 110–128.
Accad, Y., and C.L. Pekeris. 1978. Solution of the tidal equations for the M2 and S2 tides in the
    world oceans from a knowledge of the tidal potential alone. *Philosophical Transactions of the
    Royal Society of London. Series A, Mathematical and Physical Sciences* 290 (1368): 235–266.
Adams, L.H., and E.D. Williamson. 1923. Density distribution of the Earth. *Journal of the
    Washington Academy of Sciences* 13: 413–428.
Alterman, Z., Y. Eyal, and A. Merzer. 1974. On free oscillations of the earth. *Geophysical Surveys*
    1: 409–428.
Alterman, Z., and P. Rabinowitz. 1958. An interpretive system on the WEIZAC. *Bulletin:
    Mathematics and Physics* 7 (136).
Alterman, Z., H. Jarosch, and C.L. Pekeris. 1959. Oscillations of the earth. In *Proceedings of the
    Royal Society of London, Series A. 252*, 80–95. London: Royal Society Publishing. Retrieved
    from https://doi.org/10.1098/rspa.1959.0138.

Benioff, H. 1935. A linear strain seismograph. *Bulletin of the Seismological Society of America* 25: 283–309.

Benioff, H., F. Press, and S. Smith. 1961. Excitation of the free oscillations of the earth by earthquakes. *Journal of Geophysical Research* 66 (2): 605–629.

Benioff, H., B. Gutenberg, and C.F. Richter. 1953. *Progress Report.* Seismological Laboratory, California Institute of Technology.

Benioff, H., B. Gutenberg, and Richter, C.F. 1954. Seismological Laboratory Report. *Transactions, American Geophysical Union* 35 (985).

Ben-Menahem, A. (1995). A concise history of mainstream seismology: origins, legacy, and perspectives. *Bulletin of the Seismological Society of America* 85 (4): 1202–1225.

Ben-Menahem, A. 2003. Seismology and physics of the Earth's interior at the Weizmann Institute of Science, Rehovot, Israel, 1954–2000. In *International Handbook of Earthquake and Engineering Seismology,* Lee W.H., and P.C. Hiroo Kanamori, Part B, Chapter 79.31. Dordrecht: Elsevier Science.

Brush, S.G. 1996. *A History of Modern Planetary Physics: Nebulous Earth (Vol. 1).* Cambridge: Cambridge University Press.

Bullard, E.C. 1957. The density within the Earth. *Verhandlungen Der Geologische Mijnbouw Genootschap Nederlands Kalan.* 18: 23–41.

Bullen, K.E. 1940. The problem of the Earth's density variation. *Bulletin of the Seismological Society of America* 30 (3): 235–250.

Bullen, K.E. 1942. The density variation of the earth's central core. *Bulletin of the Seismological Society of America* 32 (1): 19–29.

Bullen, K.E. 1950. An Earth model based on a compressibility-pressure hypothesis. *Geophysical Journal* 6 (1): 50–59.

Bullen, K.E. 1975. *The Earth's Density.* London: Chapman and Hall.

Corry, L., and R. Leviathan. 2019. *WEIZAC: An Israeli Pioneering Adventure in Electronic Computing (1945–1963).* Berlin: Springer.

Darwin, G.H. 1879. On the bodily tides of viscous and semi-elastic spheroids, and on the ocean tides upon a yielding nucleus. *Philosophical Transactions of the Royal Society of London* 170: 1–35.

Darwin, G.H. 1882. On the stresses caused in the interior of the earth by the weight of continents and mountains. *Philosophical Transactions of the Royal Society of London* 173: 187–230.

Freeman, G. 2004. *Biographical Memories V.85.* Retrieved 03 02, 2010, from National Academies Press: http://www.nap.edu/catalog/11172.html

Helsinki. (1960). XII General Assembly of the International Union of Geodesy and Geophysics. *Geophysical Journal International* 3: 462–475.

Hestenes, D. 1999. *New Foundations for Classical Mechanics,* 2d ed. Dordrecht: Kluwer Academic.

Jarosch, H.S. 1962. *Oscillations of the Earth.* PhD Dissertation. The University of London.

Jobert, N. 1957. Sur la période propre des oscillations sphéroïdales de la Terre. *Comptes Rendus Hebdomadaires Des Séances De L'académie Des Sciences* 245 (22): 1230–1232.

Lamb, H. 1881. On the vibrations of an elastic sphere. *Proceedings of the London Mathematical Society,* s1–13 (1): 189–212.

Love, A. 1892. *A Treatise on the Mathematical Theory of Elasticity.* Cambridge: Cambridge University Press.

Love, A. 1911. *Some Problems of Geodynamics.* Cambridge: Cambridge University Press.

Lowrie, W. 2007. *Fundamentals of Geophysics,* 2nd ed. New York: Cambridge University Press.

MacDonald, G.J. 1962. Handling, reducing, and interpreting geophysical data. *Science* 138 (3546): 1276–1280.

Pekeris, C.L. 1934. An inverse boundary value problem in seismology. *Physics* 5 (10): 307–316.

Pekeris, C.L. 1941. The propagation of an SH-pulse in a layered medium. *Eos, Transactions American Geophysical Union* 22 (2): 392–393.

Pekeris, C.L. 1958. Geophysics, pure and applied. *Geophysical Journal International* 1 (3): 258.

Pekeris, C.L., Z. Alterman, and H. Jarosch. 1961. Comparison of theoretical with observed values of the periods of free oscillation of the earth. *Proceedings of the National Academy of Science* 47: 91–98.

Pekeris, C.L., and Accad, Y. 1969. Solution of Laplace's equations for the M 2 tide in the world oceans. *Philosophical Transactions of the Royal Society of London. Series A, Mathematical and Physical Sciences* 265 (1165): 413–436.

Pekeris, C.L., and H. Jarosch. 1956. *The Free Oscillations of the Earth*. Rehovot: The Weizmann Institute, Rehovot, Israel.

Pekeris, C.L. 1964. A brief history of the department of applied mathematics. *Rehovot, A Journal Published by the Weizmann Institute of Science and Yad Chaim Weizmann* 1–7.

Poisson, S. 1828. Mémoire sur l'équilibre et le mouvement des corps élastiques. In *Mémoires de l'Académie Royal des Sciences de l'Institut de France*, vol. 8, 353–570. F. Didot.

Press, F., and H. Benioff. 1958. Progress report on long period seismographs. *Geophysical Journal International* 1 (3): 208–215.

Rayleigh, L. 1885. On waves propagated along the plane surface of an elastic solid. *Proceedings of the London Mathematical Society,* s1–17 (1): 4–11.

Saint-Guily, B. 1961. Quelques études océanographiques présentées à l'Assemblée d'Helsinki. *La Houille Blanche, Revue Internationale De L'eau* 47 (sup1): 369–373.

Slichter, L.B. 1967. Earth, Free oscillation of. In *International Dictionary of Geophysics*, eds Runcorn, K., 331–343. Oxford: Pergamon Press.

Stoneley, R. 1926. The elastic yielding of the earth. *Geophysical Supplements to the Monthly Notices of the Royal Astronomical Society* 1 (7): 356–359.

Stoneley, R. 1931. On deep-focus earthquakes. *Gerlands Beiträge Zur Geophysik* 29 (3–4): 417–435.

Stoneley, R. 1961. The oscillations of the earth. *Physics and Chemistry of the Earth* 4: 239–250.

Thompson, W. 1863. On the rigidity of the Earth. *Philosophical Transactions of the Royal Society of London* 153: 573–582.

Woodhouse, J.H., and A. Deuss. 2005. Theory and observations. Earth's free oscillations. In *Treatise on Geophysics*, eds. 2d ed. Schubert and Gerald, 79–115. Dordrecht: Elsevier.

Chapter 5
Ground State of Helium

Abstract In 1958, Pekeris published an article entitled, "Ground state of two-electron atoms," which was to become his most cited work. It displayed an unusual degree of proficiency in programming and exploited to their extreme the physical capacities of WEIZAC. At the core of this contribution were a series of daunting differential equations that were solved numerically, using code written by the young Yigal Accad, who in 1973 would complete a PhD on "Ocean Tides" under Pekeris.

Keywords WEIZAC · Chaim L. Pekeris · Yigal Accad Z · Egil A. Hylleraas · Douglas R. Hartree · Vladimir A. Fock · Ground state of the helium atom · Schrödinger wave equation

Pekeris's (1958) article, "ground state of two-electron atoms," was to become his most cited work and continues to be cited up to this day.[1] This was a remarkable tour-de-force of computational dexterity, involving a thorough knowledge of the relevant theoretical and observational results in physics and of the numerical methods that had been used for solving the main problems in the discipline. It also required an unusual degree of proficiency in programming as well as the full exhaustion of the physical capacities of WEIZAC. In order to grasp the innovative aspect implied by the article and the way that WEIZAC played a crucial role in reaching the results presented in it, we must start with a brief historical account of the three-body problem as reflected in the study of microscopic particles.

5.1 From Göttingen to MIT

The problem of the two-electron helium atom played a crucial role in the development of theoretical physics and quantum chemistry in the twentieth century.[2] The initial

---

[1] Google Scholar indicates more than 1270 citations and more than 412 after 2000, including 16 in 2021 (accessed on Dec. 10, 2021).

[2] Detailed descriptions of these developments can be found in the following sources, on which we relied for our account: (Gavroglu and Simões 2011; Hylleraas 1964; Kragh 2012; Park 2009; Tanner et al. 2000, pp. 501–504).

© The Author(s), under exclusive license to Springer Nature Switzerland AG 2023   73
L. Corry and R. Leviathan, *Chaim L. Pekeris and the Art of Applying Mathematics with WEIZAC, 1955–1963*, SpringerBriefs in History of Science and Technology, https://doi.org/10.1007/978-3-031-27125-0_5

Fig. 5.1 Schematic illustration of the ground state, excited states, and ionization potential. On the value 24.47 for the ground state, see footnote 3 below

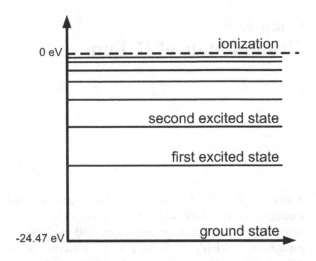

development of quantum mechanics was related to the observation that atomic emission spectra display discrete levels. Based on Max Planck's early quantum theory, the Bohr model of 1913 was highly successful in reproducing the hydrogen spectrum, but it failed when applied to two-electron atoms. Calculating the energy levels of helium became a central challenge. The physical problem in question is schematically represented in Fig. 5.1. When the system under consideration is stable, i.e., in its ground state, its energy is negative. If energy is applied (i.e., by a light-ray in a spectroscopic experiment), then successive excitation levels are achieved. The ionization potential is the energy required to reach the level where the electron is ejected from its atom. Pekeris's groundbreaking paper of 1958 comprised the calculation of the value of this potential, for the case of the helium molecule, to an unprecedented degree of accuracy.

A complete calculation of these values posed significant problems. First of all, at the purely mathematical level, a two-electron molecule is a prominent case of the three-body problem, which in the case of celestial bodies had already been acknowledged since the late nineteenth century to be intractable and to involve chaotic behaviors. But additional complexities derive from the quantum character of the problem (Barrow-Green 2008). Indeed, the helium case turned out to be unsolvable in what later became known as the "old quantum theory," where microscopic particles were still assumed to obey the laws of classical mechanics, with electrons orbiting the nucleus just like planets orbit the sun. In the "new quantum theory"—formulated by Schrödinger (1926) in terms of the wave equation and by Heisenberg (1925) in terms of the matrix formulation—microscopic particles are not taken to behave like observable objects but rather like waves. This perspective provided a first breakthrough toward solving the problem. Heisenberg applied his formalism to the helium atom, while considering wave mechanics, the electron spin (i.e., one of two types of angular momentum considered in quantum mechanics for elementary and composite particles), and the Pauli principle of exclusion (which states that no two electrons can occupy the same quantum state).

The standard formulation of the Schrödinger wave equation $\psi(x, y, z)$ for a one-electron atom with a stationary nucleus is the following:

$$\left(\frac{\partial^2}{\partial x^2} + \frac{\partial^2}{\partial y^2} + \frac{\partial^2}{\partial z^2} + 2\left(E + \frac{Z}{r}\right)\right)\psi = 0.$$

Here, Z denotes the nuclear charge, E is the energy of the system, and $r = \sqrt{x^2 + y^2 + z^2}$ is the distance of the electron to the nucleus. The lowest energy level is the ground state, and the following ones represent, respectively, the successive excited states. A very accurate calculation of the ground state of helium will be among Pekeris's important achievements in his 1958 paper.

The absolute squares of the eigenfunctions, $|\psi|^2$, are interpreted as probability distributions for the position of the particle, and hence, some restrictions need to be imposed. Specifically, the integral of $|\psi|^2$ over the entire domain should be finite (i.e., the sum of the probabilities should be 1). Under such restrictions, the eigenvalues are exactly those values of E for which the equation admits solutions. Under these restrictions, Schrödinger obtained explicit results for the eigenvalues (which are exactly those values of E for which the equation admits solutions) and for the corresponding eigenfunctions ψ.

Seen from this perspective, the equation succeeded in predicting a result similar to that previously also predicted by the old theory, but the latter did so on an ad hoc basis. Now, it was expected that the "new quantum theory," both in Heisenberg's and in Schrödinger's formulation, would allow for the prediction of new facts that the old theory could not. The task of formulating correct equations and deriving from them the energy levels that had been empirically measured for two-electron atoms became an important challenge in the field. It would be taken to imply a crucial confirmation to the new theory. It is precisely in this regard that the case of the helium atom became a focal point of departure for addressing the more general problem of many-electron atoms.

The challenge remained open for decades to come. Heisenberg succeeded in establishing a basic understanding of the spectral properties of helium, but a broad gap remained open between existing spectroscopic measurements of the ionization potential (24.4735 eV) and its numerical calculation.[3] An even broader gap remained for results obtained by Kellner (1927), within the old quantum theory (23.75 eV), and Unsöld (1927), relying on the Schrödinger Eq. (20.3 eV). The new quantum–mechanical approach helped reduce this discrepancy from about 4 to 1.5 eV, but this was still a significant amount.

---

[3] The electronvolt, eV, is defined as the amount of energy gained by the charge of a single electron when moved across an electric potential difference of one volt. According to Helgaker and Klopper (2000, p. 180), the value of 1 eV used in the older literature is different from the one accepted nowadays, so that 1 eV (old) = 1.005 eV (new). The ionization potential of helium in terms of the current value is 24.59 eV. See, e. g., https://physics.nist.gov/PhysRefData/Handbook/Tables/heliumtable1.htm. For simplicity, we use here, in each case, values as they appear in the original publications.

Various approximation methods were developed and tested for the helium atom before being applied to larger atoms. Of paramount importance in this regard was the contributions of the Norwegian physicist Egil A. Hylleraas (1898–1965), who in 1928 achieved a remarkable degree of agreement between theory and experiment for the helium problem. Working since 1926 in Göttingen with Max Born (1882–1970), Hylleraas used a trial wave function that might be interpreted as representing a first electron in an inner orbit and a second one in an outer orbit. He obtained a very good result for the ionization potential, 24.46 eV, barely 0.12 eV below the experimentally calculated value (Hylleraas 1928). At this early stage, the heavy calculations required for attaining this result were performed by Hylleraas with a rather primitive and noisy electric desk calculator known as Mercedes-Euklid. A main mathematical tool he used was the so-called Laguerre polynomials, which also became crucial for Pekeris's work, as we will see below. Relying on the classical treatise of 1924 by Courant and Hilbert, *Methoden der mathematischen Physik*, Schrödinger had already used these functions in 1926, but their actual importance was not always fully acknowledged (Mawhin and Ronveaux 2010), certainly not to the extent that Hylleraas did in his work and which Pekeris retook in 1958.

While minor, the existing discrepancy continued to bother Hylleraas. The following year, working in Oslo, he achieved an additional, major advance, which was based on replacing the angle ϑ between r_1 and r_2 in the wave equation by a new kind of coordinate, r_{12}, representing the interelectronic distance. This allowed him to introduce the electron–electron interaction directly into the function. He suggested that the energy of the ground state for the case of two electrons needs to be calculated as the absolute minimum of a variational problem formulated in terms of three independent variables, namely $s = r_1 + r_2$, $t = r_1 - r_2$, and $u = r_{12}$, where r_1, r_2 are the distances of the electrons from the nucleus, and r_{12}, as already stated, is the distance between them (Hylleraas 1929). In these terms, the Schrödinger equation is reformulated as follows[4]:

$$\nabla_1^2 \psi + \nabla_2^2 \psi + 2\left(E + \frac{Z}{r_1} + \frac{Z}{r_2} - \frac{1}{r_{12}}\right)\psi = 0, \qquad (5.1)$$

which, in the ground state and with $\psi = \psi(r_1, r_2, r_{12})$, becomes the following non-separable equation:

$$\frac{\partial^2 \psi}{\partial r_1^2} + \frac{2}{r_1}\frac{\partial \psi}{\partial r_1} + \frac{\partial^2 \psi}{\partial r_2^2} + \frac{2}{r_2}\frac{\partial \psi}{\partial r_2} + 2\frac{\partial^2 \psi}{\partial r_{12}^2} + \frac{4}{r_{12}}\frac{\partial \psi}{\partial r_{12}} + \frac{(r_1^2 - r_2^2 + r_{12}^2)}{r_1 r_{12}}\frac{\partial^2 \psi}{\partial r_1 \partial r_{12}}$$

$$+ \frac{(r_2^2 - r_1^2 + r_{12}^2)}{r_2 r_{12}}\frac{\partial^2 \psi}{\partial r_2 \partial r_{12}} + 2\left(E + \frac{Z}{r_1} + \frac{Z}{r_2} - \frac{1}{r_{12}}\right)\psi = 0. \qquad (5.2)$$

---

[4] For the sake of uniformity, and following (Pauling and Wilson 1935), we have slightly modified the notation of the following two equations.

Reducing the number of coordinates to three without actual loss of information, for a physical situation that involves in principle nine degrees of freedom, simplified the problem significantly. The justification for doing so was twofold. In the first place, the helium molecule can be analogically seen as a planetary system where a large center is taken to be static (like the Sun) and only the space coordinates, $(x_1, y_1, z_1, x_2, y_2, z_2)$, of the two very small, revolving electrons are to be determined. Now, in addition, using the three coordinates (s, t, u), instead of those six is warranted by the fact that, when considering energy, the force acting between charged particles depends only on the distance between them, and hence, the spherical symmetry underlying the problem is what allowed for this important step.

Hylleraas compared his calculations against the experimental value attained in 1924 at the Jefferson Laboratory in Harvard by Theodore Lyman (1874–1954), using photo-ionization techniques: 24.467 eV (Lyman 1924). Working out the variational problem as reformulated in terms of the three coordinates (s, t, u), Hylleraas was now able to improve over his own previous calculations, and he further improved them in a subsequent article of 1930, where he also considered the analogous atomic configurations of H^-, Li^+, and Be^{++}. He attained here the value of 24.470 eV, on which he commented (Hylleraas 1930, p. 211):

> With helium and with three additional approximations, the calculations are sharpened, so as to rule out any possibility of significant changes in the previously found value. We obtain a value for the ionization potential that surpasses the experimental one by 0.003, whereas the previous one was 0.010 eV below it. This difference is surely within the limits of experimental errors ... In the cases of H^-, He, Li^+, Be Be^{++}, the theoretical values fall all within very narrow experimental limits of error.

Hylleraas' achievements were seen at the time as a major triumph for quantum mechanics as applied to many-electron atoms and as an endorsement of the wave-mechanical equations. Nevertheless, it turned out that adapting Hylleraas's method to heavy atoms was not an easy task, as the number of terms that had to be computed increased very rapidly with the number of electrons (Pauling and Wilson 1935, pp. 220–224).

A significant technical difficulty related to Hylleraas's variational approach arose in 1935, when it was noticed that certain approximations to the wave function, based on the method of least squares, went to infinite at any of the singularities $r_1 = 0$, $r_2 = 0$, or $r_{12} = 0$. This was taken to indicate the existence of a considerable deviation of the eigenfunction in the neighborhood of these points. A possible way to attenuate this apparent problem was to use, instead of a polynomial, a power series (times an exponential factor), but it appeared that no such series, in ascending powers of r_1, r_2, r_{12}, could be a formal solution of the wave equation (Bartlett et al. 1935).

This difficulty was soon addressed in 1937 by Albert S. Coolidge (1894–1977) and his student Hubert M. James, who—back in 1933, working together at Harvard— had already obtained highly accurate values by using the Hylleraas methods and the interelectronic distance r_{12}. James and Coolidge worked with functions that could provide levels of accuracy that were better than anything known to that date by adding as many additional terms as needed (James and Coolidge 1933). The impact of their results was immediately acknowledged, but the computations grew more and more laborious and complex, as each new term that was added required the computation of additional integrals. A widely used textbook published in 1944 took due notice of this significant issue (Eyring et al. 1944, p. 217):

> [T]he labor involved in these calculations is so great even for these simple systems [such as He, H_2^+, and H_2] that it does not appear to be a profitable method of attack on molecular problems in general. Because of the mathematical difficulties involved, we are forced to use much less accurate approximations; usually, we are forced to write the wave function as some linear combination of one-electron wave functions. Although these will not give satisfactory quantitative results, they should in general be qualitatively correct and should enable us to correlate experimental chemical facts.

The tension between seeking after more accurate results and requiring prohibitive calculational capacities was mounting increasingly high, and the ensuing gap would only be closed once the automatic electronic computers became widely available. But even before that, it was also necessary to improve the numerical techniques that had been elaborated over the two preceding decades. Now, in 1937, James and Coolidge addressed the difficulty that apparently affected the Hylleraas methods by introducing an algorithmic improvement that would also prove crucial for Pekeris's attack on the problem using WEIZAC in 1958. They showed that the apparent impossibility of using polynomials to represent the function as closely as desired could be easily overcome, and that if a formal solution of the wave equation for helium did exist, then the energy given by the Hylleraas method would converge upon the correct energy. In fact, even if there exists no formal solution of the wave equation, then there is a lower bound to the energy which can be computed with any function, and the Hylleraas method will converge anyway upon this bound. The approach of Coolidge and James was based on the introduction of a linear change of variables to what later became known as "perimetric coordinates" (Coolidge and James 1937). These are defined as follows:

$$x = \delta(r_1 + r_2 - r_{12}); \ y = \delta(r_2 + r_{12} - r_1); \ z = \delta(r_1 + r_{12} - r_2),$$

and it can be shown that expressed in these terms, the restriction on the square of the eigenvalues, $\int \psi^2 d\tau = 1$, can be more flexibly handled. Below in Sect. 5.2, we describe these coordinates and their properties in greater detail, by referring to the way that Pekeris used them in his calculations.

An alternative approach to calculating values related to the ground state of helium, which deserves being mentioned here, is the one developed by Douglas R. Hartree, whom we already mentioned above. The method, known as the self-consistent-field (SCF) method, was less accurate than Hylleraas's, but on the other hand, it was much more manageable and more reliable. Hartree started to work out this approach in 1928, and his basic idea was to approach the Schrödinger equation for a multielectron atom as if it could be separated into a set of equations, one for each electron, so that each of these would be solved separately. In order to do so, one had to neglect the explicit interactions between the electrons. These interactions, however, have a real physical meaning, which is at the heart of the difficulty posed by the many-body problem. He suggested considering each electron as ideally moving in a sort of "average" field produced by the others and seeking for numerical solutions that are "self-consistent" (Gavroglu and Simões 2011, pp. 138–158).

The meaning of this is that the solution obtained when solving the equations with the correct effective potential should be the same if one calculated them in reverse. Thus, the procedure starts with an initial guess of the values of the said "average field," and it then follows an iterative process based on calculating the numerical solution of ODEs related to the wave functions. This yields a distribution of electric charge, with values that are likely to be different from the first guessed estimate. The calculation is repeated until the two processes of finding the wave functions and their electric fields are mutually consistent. The convergence of this process of successive approximations, however, is not guaranteed beforehand. The field initially chosen and the expansion chosen for the functions are crucial. If the chosen field remains unchanged through another round of calculation, and a final field is found which is the same as the initial one, then a "self-consistent" field is found. This field, moreover, is finally taken to be characteristic of the atom under consideration (Hartree 1928a, b, c, 1929).

As originally formulated, Hartree's method was soon met with criticism waged at him independently from various directions (Park 2009, pp. 50–58). To be sure, he had justified his method only by reference to qualitative arguments and based on the agreement of calculation with experiment. He had not at all considered some of the most basic issues related to many-electron systems such as the Pauli exclusion principle, the spin state of electrons, or the Heisenberg resonance phenomenon (i.e., the exchange of energy between states described by the same eigenfunction series). Thus, of particular importance was the criticism raised by John Clarke Slater (1900–1976), then at Harvard and MIT, and by Vladimir A. Fock (1898–1974), at Leningrad. Initially, Slater had compared the values obtained by Hartree as a convenient way to assess the accuracy of his own computations of the diamagnetic susceptibility of helium and of the forces of repulsion between two normal helium atoms. He thus wrote:

> The wave function found in the present paper is more complicated than Hartree's in the matter of the way in which it takes the interaction energy between electrons into account. But they can be computed equally well from either method, and this permits a comparison of the present results with Hartree's. The discrepancies between the two are nowhere greater than one or two percent. This is highly satisfactory, both in that it verifies the present method and Hartree's, and also that it justifies us in believing this density distribution to be correct within a narrow limit of error. The other numerical results are also gratifying: the diamagnetic susceptibility agrees with the experimental value within a percent and a half, and the results for collisions of two atoms agree with experiment within the rather wide limit caused by uncertainties in the kinetic theory treatment of the imperfect gas problem. (Slater 1928)

Later on, however, he indicated that Hartree had overlooked the principle of anti-symmetry of the wave function, which states that this function needs to be—in the case of a many-electron atom—anti-symmetric with respect to interchange of coordinates between any two electrons. This problem, however, could be solved in a relatively simple manner by adopting what came to known as the "Slater determinant," in which the wave function, as originally introduced by Heisenberg and Paul Dirac in 1926, is represented in a way that satisfies inherently the anti-symmetry requirement, e.g., $\psi(x_1, x_2) = -\psi(x_2, x_1)$. Notice that in this formulation, the coordinates are vectors having not only three spatial degrees of freedom but also an intrinsic spin coordinate. Thus, the spatial part can itself be either symmetric or anti-symmetric, but when multiplied, respectively, by either an anti-symmetric or a symmetric spin part, the total wave function will remain anti-symmetric. In terms of the mathematical representation, the required anti-symmetry results from the fact that switching two particles means switching two rows in the determinant, thus changing its sign. Moreover, the very use of such a determinant is itself a warranty that two particles cannot be at the same quantum state, as this would result in two equal rows and hence the value zero for the determinant. These important issues will reappear in Pekeris's treatment of the helium atom, as will be seen below in Sect. 1.1.1.

Fock's criticism was similar, and he generalized Hartree's method so that it would comprise additional dynamic variables, such as the exchange forces. The method he introduced in 1930 relied strongly on group theoretical tools and was slow in gaining acceptance among contemporary physicists (Fock 1930). More generally speaking, the kinds of computational methods developed by these researchers were not seen in the discipline with the same kind of enthusiasm aroused by more purely theoretical works developed at the time, despite the difficulties encountered in the attempts to prove definitive results using the latter. Against the background of such critical reactions, Hartree in 1935 reformulated his method to be more suitable for the purposes of specific numerical calculations. Referring to other existing methods, he wrote:

> The principles and general theory of a method for including exchange effects have already been given some years ago by Fock; they were suggested independently about the same time by Slater. But the practical handling of Fock's equations is a problem of numerical technique altogether more complex than the solution of the self-consistent field problem without exchange, and, as far as we are aware, the only complete solution yet obtained for any case in which the main features of the problem are fully shown is that of sodium, for which the results, and methods used to obtain them have quite recently been given by Fock and Petrashen. The solution of Fock's equations given in the present paper had been completed, and the subsidiary calculations of energy values were in progress, when this paper appeared. The methods we have used differ considerably from those of Fock and Petrashen and appear simpler for practical work; for example, we make no use of analytical approximations which are usually not so convenient to handle numerically as they might appear, or of the Green's function constructed by them. (Hartree and Hartree 1935)[5]

What set Hartree apart from most of his colleagues was his awareness of the technical difficulties that could be encountered in designing and actually implementing any of the numerical methods proposed for calculations. In his original method, for each cycle of approximations and for each of the electrons in atoms, he had to find the solution of a differential equation that could not be solved analytically. He developed a technique of numerical integration specifically designed to solve the equation at

---

[5] See also (Schweber 1990).

numerous fixed points of r, the distance from the nucleus. He was fully absorbed by the computations to the extent that sometimes he failed to keep abreast of the recent development of quantum theory. In replying to Slater's criticism, for example, he wrote:

> Some of the steps were not clear to me without a bit of work and looking up the general theory, but that is my fault; my time has been so taken up with the development of the numerical technique of evaluating the self-consistent field, and with the actual computing of particular cases, that I am not as familiar as I should be with the general theory outside what I have required for my work, which is not much.[6]

Throughout the late 1920s and 1930s, his unique expertise in numerical analysis was used to develop ingenious approximation methods for evaluating in an efficient manner the self-consistent fields of atoms of increasing atomic numbers. The Hartree–Fock method required the use of much larger computational resources than those required by Hartree's earlier method or by existing empirical models. Initially, their method was applied only for atoms featuring spherical symmetry that allowed a considerable simplification of the problem. Still, calculating solutions of the Hartree–Fock equations for a medium-sized atom was an extremely laborious task and even small molecules required computational resources that were unrealistic if performed manually. Thus, also in the case of Hartree–Fock method, it was not until the introduction of electronic computers that it could become widely used.

It is worth indicating that much of the actual computations appearing in Hartree's publications in the 1920s and 1930s were performed by his father with the help of a desk calculating, handle-operated machine of the Brunsviga type, with metal levers for setting the numbers, and with a bell that rang in cases of overflow. In 1930, he heard from Slater—who had just moved to MIT as head of the physics department and gathered around him several theoreticians interested in atomic wave functions—about Bush's differential analyzer, then still in development (Owens 1986). This analog machine was specifically designed for solving differential or integral equations, and it was particularly able to handle efficiently one-dimensional wave equations. Slater planned to use this machine for implementing Hartree's algorithm of self-consistent fields, or some similar to it. Being an analog machine, the differential analyzer could provide for solutions to problems for which the existing numerical methods were cumbersome and became increasingly laborious the more complicated the equations grew. On the other hand, the mechanical nature of the machine imposed serious limitations on the speed and accuracy of its operation.

As already indicated, Hartree, back from a visit to MIT in 1933, built up in Manchester a Meccano model of the analyzer, more as an amusement than with serious scientific purposes in mind. But as some preliminary results became more successful than expected, he devised the construction of a more professional, full-scale model and contracted the Metropolitan-Vickers Electrical Company to do the job, which was completed in 1935 (Hartree and Porter 1935). At roughly the same time, a Central Mathematics Laboratory was established at Cambridge mainly at the initiative of chemists intent on solving problems related to electronic structure theory.

---

[6] Cited in (Park 2009, p. 59).

They worked with an existing, rich collection of mathematical tables, Brunsviga machines, a differential analyzer, and some minor special-purpose machines (Wilson 1997, pp. 88–89).

Side by side with these advances in machine building and its sustained application in the discipline, the current theoretical advances were yielding increasingly complicated wave equations, which underscored the computational limitations of the existing machines and methods. The Metropolitan-Vickers differential analyzer, for example, was never actually used for its original purpose, though it still contributed to important tasks in the industry and the government as well as to other scientific tasks. But Hartree was keen in stressing the role of machines as an aid to the calculations on which much preliminary thought must be invested. In his 1947 Inaugural Talk at Cambridge, he stressed that the iterative methods that he had developed since his 1928 paper were very suitable for mechanization. At the same time, however, he cautioned that:

> Use of the machine is no substitute for the thought of organizing the computations, only for the labor of carrying them out. This point seems to me of great importance and to be missed entirely by those who speak of a machine of this kind as an "electronic brain." Even if the organization of the calculations is done by the machine, as is possible in future developments, the operator will still have to think about the sequence of operating instructions which will enable the machine to do this organization. (Hartree 1947)

But as in many other fields of research, the advent of the electronic computer turned out to be the most significant watershed that changed the very idea of what is worth computing in the theory of electronic and molecular structure. The first digital electronic computer used at MIT, for example, for solving computational tasks in the discipline was the Whirlwind, which became operational in 1951. Still, real change did not happen overnight. It required several years of hands-on experience on the side of the researchers as well the development of high-level languages and reusable routines for specific tasks such as calculating definite integrals or solving systems of linear equations.[7]

On April 30, 1958, Robert Mulliken (1896–1986) and Clemens C. Roothaan (1918–2019) presented to the National Academy of Sciences a paper entitled "Broken Bottlenecks and the Future of Molecular Quantum Mechanics." They enthusiastically spoke about "the future application of quantum mechanics to chemistry and molecular and solid-state physics" and compared it with the application by the likes of Lagrange and Hamilton of the principles of classical mechanics in the years following the discovery of Newton's laws of motion. At the same time, however, they also indicated the main obstacle currently preventing this enormous potential from materializing: quantum mechanics, they said, had shown itself "extremely fruitful for a qualitative theoretical explanation of, for instance, the empirical rules of valence. But until very recently, it has for the most part been less successful in quantitative predictions." Among the main reasons for this, they indicated the very high "complexity in terms of the mathematical analysis and the computational efforts required

---

[7] For a survey of nearly contemporary methods for solving systems of linear equations, see (Forsythe 1953).

for quantitative results" arising from the more difficult problems of chemistry and molecular physics. The increasing availability of automatic electronic computers, however, allowed predicting with confidence "colossal rewards" and a bright future for large-scale quantum–mechanical calculations of the structure of matter:

> [M]achine calculations will lead gradually toward a really fundamental quantitative under-standing of the rules of valence theory and the exceptions to these; toward a real understanding of the dimensions and detailed structures, force constants, dipole moments, ionization poten-tials, and other properties of stable molecules and equally of unstable radicals, anions, and cations, and chemical reaction intermediates; toward a basic understanding of activated states in chemical reactions, and of triplet and other excited states which are important in combustion and explosion processes and in photochemistry and in radiation chemistry; also of intermolecular forces; further, of the structure and stability of metals and other solids; of those parts of molecular wave functions which are important in nuclear magnetic reso-nance, nuclear quadrupole coupling, and other interactions involving electrons and nuclei; and of very many other aspects of the structure of matter which are now understood only qualitatively or semiempirically. (Mulliken and Roothaan 1959, p. 397)

Thus, in the view of these two leading chemists, intensive calculations such as afforded by the new electronic computers would lead not only to obtaining important quantitative values, but also to a *deeper theoretical understanding* of some of the most significant issues of the discipline. And this was precisely the time when Pekeris was hard at work with WEIZAC, successfully completing some of the most complex computational tasks that the Hylleraas approach had opened, but not theretofore fully achieved, concerning the ground state of the two-electron atoms.

5.2 Pekeris and the Helium Atom

In the late 1950s, when he started to work on the question of the helium atom while calculating with WEIZAC, Pekeris was well aware of the fundamental scientific issues and current challenges—both theoretical and computational—that attracted the attention of the practitioners in the discipline. He was also aware of the exper-imental results known at the time. Like in other fields, even before his arrival in Israel, Pekeris had made explicit his intention to use the electronic computer to solve problems in quantum mechanics, involving calculations that were theretofore beyond the capacity of large and highly trained teams of human computers. In the above-mentioned letter of 1946 to Getzoff, Pekeris wrote:

> These machines can perform calculations in one hour which would ordinarily task hundreds of computers for many weary months. They also make feasible certain analyzes which hitherto have not been attempted because of their complexity. A case in point ... is the application of quantum mechanics to chemical problems. This field is practically virgin today because only the simplest atoms have thus far yielded to unaided human analysis.[8]

Testimonies of his continued interest in this problem appeared scattered throughout his publications. In 1950, he published a note where he listed some open

---

[8] See above, footnote 11.

questions regarding the behavior of helium at low temperatures and the attempts made to address the specific issues that arise from it (Pekeris 1950). His landmark contribution in this field appeared in an article of 1958, whose contents we proceed to explain now in some detail, with a view to describing how the use of WEIZAC played a fundamental role in contributing to this important branch of twentieth century science.

By the late 1950s, while some progress had been attained with numerical solutions, the methods derived from Hylleraas's work continued to be the current practice for finding exact solutions related to the question of the helium atom. They were not beyond criticism, however, and Pekeris was well aware of this. In fact, in a review of 1960 on work done with WEIZAC, Pekeris described his decision to take up this research in the following words:

> The [solution of the Schrödinger equation] has attracted attention in recent years in connection with the determination of the magnitude of the Lamb shift in Helium, a program initiated by Herzberg in collaboration with Chandrasekhar. I thought of the early stages of their program the other day when von Kármán lifted the phone at the Dan Hotel and asked to talk to the manager. "Are you under the illusion that you are running a first-class hotel?" he asked. "I am sorry to report that the bathroom light is on the blink." In two minutes, this was fixed. Essentially in 1953, Chandrasekhar turned to the physicists and asked: "are you under the illusion that Hylleraas' ionization energy for Helium is correct to within 1 cm^{-1}? If so, you should know that I have just discovered an error of 21.5 cm^{-1}."[9]

Pekeris was referring to a recent article by the two future Nobel-Prize winners, Gerhard Herzberg (1904–1999) and Subrahmanyan Chandrasekhar (1910–1995), (Chandrasekhar et al. 1953),[10] where they discussed a subtle yet important phenomenon discovered by yet another future Nobel-Prize awardee, Willis Lamb (1913–2008), working with his student Robert Curtis Retherford (1912–1981) (Lamb and Retherford 1947). The Lamb Shift, as the phenomenon is known, refers to a slight difference (a shift in the spectral lines) between two energy levels of the hydrogen atom, which—according to the Schrödinger equation—should be equal, as in theory both depend only on the same principal quantum number n, arising from the radial part of the equation (Lamb 1951). While truly small, this difference presented a real challenge to the theory and attracted much attention at the time, becoming a central challenge for many leading physicists of the generation (Cohen et al. 2009, pp. 6–7). While attempting to improve the precision of the known spectroscopic determinations of the ionization potential of helium, Chandrasekhar, Elbert, and Herzberg found several inaccuracies in the current literature and indicated the existence of an unexplained additional shift of 21.5 cm^{-1}.[11]

---

[9] Pekeris, "Review of the Research Carried Out on WEIZAC. Symposium in Applied Mathematics. Weizmann Institute, April 26, 1960," p. 4 (WIA 227 -1/74).

[10] Donna Elbert was Chandrasekhar's assistant who did the calculations using a Marchant mechanical calculator (Donnelly 2011).

[11] Here, the unit used for measuring energy is the "inverse cm," which is proportional to the wavenumber and frequency and which chemists typically find convenient for their work. In current terms, 1 eV = 8065.54 cm^{-1}. See http://doi.org/10.5281/zenodo.22826 (accessed Dec 08, 2021).

Still, Pekeris continued to adopt Hylleraas's methods, which he dubbed "classical," as the starting point of his work, while declaring, he had developed a new method for treating the threeaccount the Lambe-body problem in quantum mechanics, which was particularly suited to the electronic computer. As will be seen below, he did not at all ignore the need to take into Shift.

As already indicated, a basic idea of Hylleraas's approach related to the possibility of formulating the equation in terms of only three variables, r_1, r_2, r_{12}, whereby, in the ground state, $\psi(x_1, y_1, z_1, x_2, y_2, z_2) = \psi(r_1, r_2, r_{12})$. But because of the triangular conditions, $r_1 + r_2 \geq r_{12} \geq |r_1 - r_2|$, it turns out that these variables cannot be independent. Hence, Pekeris introduced an adapted version of the perimetric coordinates of James and Coolidge, as follows:

$$u = \epsilon(r_2 + r_{12} - r_1),$$
$$v = \epsilon(r_1 + r_{12} - r_2),$$
$$w = 2\epsilon(r_1 + r_2 - r_{12}).$$

This is a very important point in Pekeris's treatment, because the new coordinates, such as u, v, and w, are *not* constrained by the triangular inequality, and hence, they are independent. In addition, they range each from 0 to ∞, so that the domain of the coordinates is a rectangular region of the space, thus considerably simplifying the calculations with the integrals that arise from handling the quantum configuration of the three-particle system (Frolov 2006, pp. 15422–15424). The added parameter, $\epsilon = \sqrt{-E}$ (E being the energy of the system), is introduced in order to attain a more accurate approximation when using a finite number of terms, at the cost of sacrificing the actual asymptotic behavior of ψ. This behavior is related to the form that the wave function is assumed to have in the Hylleraas method with perimetric coordinates, namely.

$$\psi = \exp\left[-\frac{k(r_1 + r_2)}{2}\right] \times \sum_{l,m,n=0}^{\infty} c_{lmn} k^{l+m+n} (r_1 + r_2)^l (r_1 - r_2)^m (r_{12})^n.$$

Here, the "scale parameter," k, and the coefficients, c_{lmn}, are derived as part of the determination of the variational form of Eq. (5.2). For this latter equation to behave asymptotically, it is necessary that in its *exact* solution, for large values of $r_1 + r_2$, k be equal to 2ϵ. But within the *variational* approach, k may be allowed to deviate from 2ϵ, thus achieving the desired behavior, which is more adequate for numerically calculating the approximate value, and still to do so with great accuracy.

In these terms, the function can be now written in parametric coordinates as follows:

$$\psi = \exp\left[-\frac{u + v + w}{2}\right] \times F(u, v, w), \tag{5.3}$$

$$\epsilon\{(4u^2v+4uv^2+4u^2w+4uvw+2uw^2)F_{uu}$$
$$+(4u^2v+4uv^2+4v^2w+4uvw+2vw^2)F_{vv}$$
$$+(8u^2w+8v^2w+4uw^2+4vw^2)F_{ww}$$
$$-4uw(2u+w)F_{uw}-4vw(2v+w)F_{vw}$$
$$+(-4u^2+4v^2+2w^2+4uw+4vw+8uv-4uvw$$
$$-4uv^2-4u^2v)F_u+(4u^2-4v^2+2w^2$$
$$+4uw+4vw+8uv-4uvw-4uv^2-4u^2v)F_v$$
$$+(8u^2+8v^2-4w^2-2uw^2-2vw^2-4u^2w-4v^2w)F_w$$
$$-4(u+v)(u+v+w)F\}$$
$$+[4Z(u+v)(u+v+w)-(2u+w)(2v+w)]F=0.$$

where F is the function to be determined. By very laborious calculations, Pekeris derived the explicit, and rather daunting, expression for the linear partial differential equation satisfied by F. Figure 5.2. reproduces this expression as it appears in the original article.

This expression has also a variational equivalent (not shown here) which, proceeding along the lines of Hylleraas, allowed Pekeris to assume that the function F may be written in terms of Laguerre polynomials, $L_n(w)$,

$$L_n(w) = \sum_{k=0}^{n}\binom{n}{k}\frac{(-w)^k}{k!},$$

where

$$\int_0^\infty e^{-w}[L_n(w)]^2 dw = 1.$$

F would be expressed, in these terms, as an expansion whose coefficients $A(l, m, n)$ needed to be determined. The expansion was written as follows:

$$F = \sum_{l,m,n=0}^{\infty} A(l, m, n)L_l(u)L_m(v)L_n(w), \tag{5.4}$$

Now, the Laguerre polynomials $L_n(x)$ are known to satisfy, together with $L'_n(x)$, $L''_n(x)$ (their first- and second-order derivatives with respect to x), the following basic relations[12]:

(1) a differential equation of the second order,

$$xL''_n(x) = (x-1)L'_n(x) - nL_n(x);$$

---

[12] Pekeris was relying on the treatment appearing in (Erdelyi et al. 1953).

$4(l+1)(l+2)[-Z+\epsilon(1+m+n)]A(l+2,m,n)+4(m+1)(m+2)[-Z+\epsilon(1+l+n)]A(l,m+2,n)$

$\quad+4(l+1)(m+1)[1-2Z+\epsilon(2+l+m)]A(l+1,m+1,n)+2(l+1)(n+1)[1-2Z+\epsilon(2+2m+n)]$

$\quad\times A(l+1,m,n+1)+2(m+1)(n+1)[1-2Z+\epsilon(2+2l+n)]A(l,m+1,n+1)+(n+1)(n+2)A(l,m,n+2)$

$\quad+(l+1)\{4Z(4l+4m+2n+7)-8m-4n-6-2\epsilon[(m+n)(4m+12l)+n^2+12l+18m+15n+14]\}A(l+1,m,n)$

$\quad+(m+1)\{4Z(4l+4m+2n+7)-8l-4n-6-2\epsilon[(l+n)(4l+12m)+n^2+12m+18l+15n+14]\}A(l,m+1,n)$

$\quad+4(n+1)\{Z(2l+2m+2)-l-m-n-2-\epsilon[-l^2-m^2+4lm+2ln+2nm+3l+3m+2n+2]\}A(l,m,n+1)$

$\quad+4\epsilon(m+1)(m+2)nA(l,m+2,n-1)+4\epsilon(l+1)(l+2)nA(l+2,m,n-1)$

$\quad+2\epsilon l(n+1)(n+2)A(l-1,m,n+2)+2\epsilon m(n+1)(n+2)A(l,m-1,n+2)$

$\quad+\{4(2l+1)(2m+1)+4(2n+1)(l+m+1)+6n^2+6n+2-4Z[(l+m)(6l+6m+4n+12)-4lm+4n+8]$

$\quad+4\epsilon[(l+m)(10lm+10mn+10ln+10l+10m+18n+4n^2+16)+lm(4-12n)+8+12n+4n^2]\}A(l,m,n)$

$\quad+4l(m+1)[1-2Z+\epsilon(1+l+m)]A(l-1,m+1,n)+4(l+1)m[1-2Z+\epsilon(1+l+m)]A(l+1,m-1,n)$

$\quad+2l(n+1)[1-2Z+\epsilon(2m-4l-n)]A(l-1,m,n+1)+2m(n+1)[1-2Z+\epsilon(2l-4m-n)]A(l,m-1,n+1)$

$\quad+2(l+1)n[1-2Z+\epsilon(2m-4l-n-3)]A(l+1,m,n-1)+2(m+1)n[1-2Z+\epsilon(2l-4m-n-3)]A(l,m+1,n-1)$

$\quad+2l\{-(4m+2n+3)+Z(8l+8m+4n+6)-\epsilon[(m+n+1)(12l+4m+2)+n+n^2]\}A(l-1,m,n)$

$\quad+2m\{-(4l+2n+3)+Z(8l+8m+4n+6)-\epsilon[(l+n+1)(12m+4l+2)+n+n^2]\}A(l,m-1,n)$

$\quad+4n\{-(l+m+n+1)+Z(2l+2m+2)-\epsilon[(l+m)(1+2n-l-m)+6lm+2n]\}A(l,m,n-1)$

$\quad+2\epsilon n(n-1)(l+1)A(l+1,m,n-2)+2\epsilon n(n-1)(m+1)A(l,m+1,n-2)+4\epsilon l(l-1)(n+1)A(l-2,m,n+1)$

$\quad+4\epsilon m(m-1)(n+1)A(l,m-2,n+1)+4l(l-1)[-Z+\epsilon(1+m+n)]A(l-2,m,n)$

$\quad+4m(m-1)[-Z+\epsilon(1+l+n)]A(l,m-2,n)+n(n-1)A(l,m,n-2)$

$\quad+4lm[1-2Z+\epsilon(l+m)]A(l-1,m-1,n)+2ln[1-2Z+\epsilon(2m+n+1)]A(l-1,m,n-1)$

$$+2mn[1-2Z+\epsilon(2l+n+1)]A(l,m-1,n-1)=0. \quad (22)$$

Fig. 5.3 33-term "monster equation" at the heart of Pekeris' 1958 computation. Reproduced from (Pekeris 1958, p. 1650). Copyright by the American Physical Society

(2) a differential equation of the first order,

$$xL'_n(x) = nL_n(x) - nL_{n-1}(x);$$

(3) a recurrence relation,

$$xL_n(x) = -(n+1)L_{n+1}(x) + (2n+1)L_n(x) - nL_{n-1}(x).$$

Using these relations, Pekeris was able to get rid of all differentiations as well as of any monomials involving u, v, w. By now substituting Eq. (5.4) into the explicit expression for the linear PDE satisfied by F (Fig. 5.2), Pekeris derived (again, purely by hand, and by tedious and exacting labor) an explicit recursion expression for the coefficients $A(l, m, n)$. For each of these coefficients, he obtained an even much more daunting (in fact *monster*[13]), 33-term differential equation. This monster equation is shown here in Fig. 5.3, reproduced from the original article.

Pekeris indicated that this "recursion relation does not look attractive" (and one cannot but agree with him), but he also stressed that it has the very strong advantage "that all the coefficients in the resulting determinant are integers, which facilitates storage in the computer" (Pekeris 1972). In raising this important point, Pekeris surely knew what he was talking about. A fundamental challenge arose in the early stages of implementing electronic methods for scientific calculations related to the issue of

---

[13] A most adequate expression used in (Koutschan and Zeilberger 2011, p. 54), which we willingly adopt here.

representing real numbers, possibly involving an infinite decimal expansion, with a finite number of bits. The method of *floating-point* representation, adopted from very early on, derived from the idea of "scientific decimal notation," comprising a *fractional part (mantissa)* and an *exponent*.[14] Early electronic computers, however, were "fixed-point machines," and typically did not support floating-point representation at the hardware level, and implementing the representation required the use of considerable memory resources. In their seminal 1946 design document, for example, Burks, Goldstine, and von Neumann explicitly spoke about this dilemma and were quite hesitant when considering "whether the modest advantages of a floating binary point offset the loss of memory capacity and the increased complexity of the arithmetic and control circuits" (Burks et al. 1946). Also, Wilkes referred to this important point retrospectively when speaking about the design principles of the EDSAC (Wilkes 1997):

> It went without saying that the EDSAC would be fixed-point machine. Floating-point relay machines had been built, and the merits of floating-point operation were fully recognized. However, it would be long time before any electronic engineer would feel sufficiently confident with vacuum tube technology to attempt the design of floating-point machine.

It was only in 1956, with the introduction of the IBM 704, that a floating-point arithmetic hardware was widely adopted. According to Ceruzzi (2000), this was the main market leverage offered by this specific machine.

In the absence of a floating-point hardware system, the programmer had two options: either to use floating-point software, or to use the fixed-point arithmetic directly. In the second case, as Wilkes formulated it, "scaling is the name of the game" (Wilkes 1997). The meaning of this was that when calculating a given formula, it had to be rewritten with all of its variables scaled to fit within the same range. Scaling was a major challenge for the programmer, as it might lead to significant digit loss and, consequently, accumulated rounding-off errors. Various techniques of "floating-point software" could be used as an alternative that would alleviate such problems, but Rabinowitz, as well as many other programmers whom Pekeris consulted, believed that the cost of this approach was "much more time spent in computation and less storage capacity left for data" (Davis and Rabinowitz 1954).

5.2.1   Helium Equations with WEIZAC

Pekeris had thus formulated recursion relations, where each of the equations involved is linear on the two main physical magnitudes Z, ϵ, and where ϵ is to be found. In each case, the thirty three terms correspond to shifted indexes for $A(l, m, n)$, namely $A(l + \alpha, m + \beta, n + \gamma)$ (α, β, γ taking values between -2 and 2). Each of these terms, as they show up in the equation, is multiplied by a polynomial expression of

---

[14] See, for example, (Goldberg 1991). (Kornerupi and Matula 2010, p. Chap. 7) give a comprehensive discussion on floating-point number system.

the indexes l, m, n of up to degree 3. But Pekeris further noticed that these lengthy equations could, in fact, be reformulated in the following terms:

$$\sum_{\alpha,\beta,\gamma=-2}^{+2} C_{\alpha,\beta,\gamma}(l, m, n) A(l + \alpha, m + \beta, n + \gamma) = 0,$$

in which case, we also have

$$C_{-\alpha,-\beta,-\gamma}(l, m, n) = C_{\alpha,\beta,\gamma}(l - \alpha, m - \beta, n - \gamma).$$

Recall now that from the work of Hartree and Fock, and from the introduction of the Slater determinant, the requirement of anti-symmetry for the wave function had been successfully met.

Notice also that if ψ is symmetrical in the two electrons, then

$$A(l, m, n) = A(m, l, n),$$

whereas if ψ is anti-symmetrical, then

$$A(l, m, n) = -A(m, l, n)$$

These two cases correspond, respectively, to the so-called para and ortho states of the helium atom, which refer to the way the spins of the two electrons couple with each other. In the ortho state, the two electrons have parallel spins, whereas in the para state, they have opposite spins. In the early days of the new quantum mechanics, Heisenberg and Friedrich Hund had introduced this separation (Heisenberg 1927; Hund 1927), showing that the differences between the states, in the case of the hydrogen molecule, are due to the dependence of the exchange energy on their symmetry properties (Bloch 1976; Sancho and Plaja 2008). In the case of the helium atom, the ground state can only be para, because in the ortho state, both the orbital momentum and the spin are not equal to zero, which would produce a highly reactive situation in which the helium atom would behave similarly to the hydrogen atom. This additional condition allowed Pekeris to further reduce by about half the number of equations to be solved.

Thus stated, each 33-term equation represents a set of an infinite (actually ∞^3) number of homogenous linear equations on an infinite (actually ∞^3) number of unknowns. In general, such equations have no non-trivial solutions for the $A(l, m, n)$, but when the determinant vanishes the eigenvalues of the parameter ϵ which are obtained, and there are solutions. More importantly, the largest such eigenvalue corresponds to the ground-state energy of the atom, which is here at the focus of the physical problem investigated.

Table 5.1 Pekeris' scheme for reducing the scope and complexity of the calculations related to the "monster equation"

l	M	n	ω	K	L	m	n	ω	K
0	0	0	0	1	1	1	1	3	11
0	0	1	1	2	0	3	0	3	12
0	1	0	1	3	1	2	0	3	13
0	0	2	2	4	0	0	4	4	14
0	1	1	2	5	0	1	3	4	15
0	2	0	2	6	0	2	2	4	16
1	1	0	2	7	1	2	2	4	17
0	0	3	3	8	0	3	1	4	18
0	1	2	3	9	1	2	1	4	19
0	2	1	3	10	0	4	0	4	20

Now, to further reduce the scope of the problem and make it more manageable, Pekeris considered only values $l, m, n \geq 0$, for which $l + m + n \leq \omega$, for some finite ω, and $l \leq m$, while setting all the $A(l, m, n) = 0$, whenever $l + m + n > \omega$. Moreover, in order to consider the possibility of handling this problem algorithmically, it was necessary for Pekeris to be able to order the triplets l, m, n, while assigning each of them an index k, which he did with the help of an ingenious scheme. In the symmetrical case, for example, he obtained the ordering indicated in Table 5.1.

If the triplets are successively substituted in the monster equation via the ordering provided by k, while setting $A(l, m, n) = B_k$, then the system of equations becomes

$$\sum_k C_{ik} B_k = 0, \tag{5.5}$$

where $C_{ik} = a_{ik} + \epsilon b_{ik}$, with a_{ik}, b_{ik}, integers and $C_{ik} = C_{ki}$.

It is remarkable—and worth stressing once again—the extent to which the entire approach followed by Pekeris focused on its possible, efficient implementation in an electronic computer. Thus, for example, he indicated that as his method was biased to attaining a proper behavior at infinity, relying on Eq. (5.3) implied giving up the "scale parameter k," and hence in Eq. (5.5), one obtains a value of 2ε, which might be less accurate than that which one would obtain if k was preserved. This price, however, was worth paying because of reasons related to the implementation in WEIZAC. For one thing, because the coefficients a_{ik}, b_{ik}, of the determinant arising from Eq. (5.5) are all integers. For another thing, because the resulting determinant is very sparse, having on average only about 20 non-vanishing terms per row.

Pekeris knew his machine and its capabilities very well. He explained that since the determinant contains only integers, they could be handled with double-precision accuracy (i.e., 18 decimals), and still, four times as many elements of the determinant could be accommodated in the fast memory.[15] Furthermore, because of the sparsity of the determinants, the capacity of the fast memory of WEIZAC would not overflow for any determinant of up to the order of 214. In the determinants obtained for the Hylleraas-type of solution, by contrast, every element was non-zero and not an integer. Such determinants could be accommodated for calculation with WEIZAC only up to an order of about 42.

But it was not just a matter of knowing WEIZAC, its strengths and weaknesses. Pekeris's accumulated experience in computation-intensive projects, going way back to his days at the Division of War Research in Columbia, was directly reflected in this work as well. Thus, for instance, an early version of the iteration process that Pekeris applied here for calculating the values of the determinants and the corresponding eigenvalues had already been introduced in an article of 1946 dealing with perturbations in processes of propagation of microwaves (Pekeris 1946). Back then, the calculations were performed by a group of six human computers, and the new method introduced in the article was meant to overcome the limitations of previously applied methods that became extremely complex even after very few iterations. Now, the iteration process was used for solving the determinant with terms a_{ik}, b_{ik}, arising in Eq. (5.5). The program that computed the values of the coefficients and then performed the iterations, following the ordering of triplets $l, m, n,$ as indicated above in Table 5.1, was written and run in WEIZAC by the young Yigal Accad (b. 1936). Accad, who in 1973, would complete a PhD on "Ocean Tides" under Pekeris, had at the time just completed his military service, and was drafted to the job by Pekeris, thereby starting a years-long collaboration between the two.

The basic principle of the iteration was to truncate the determinant at the order n, setting $B_1 = 1$, and then using the n equations to solve for the remaining n-1 values of B_n, as well as of ϵ. The values obtained for n-1 were used as the initial values for the order n, together with $B_n = 0$. Once the successive values of ϵ stabilized to a predefined degree of accuracy, the iterations were stopped. The first round of iterations started with $n = 1$, proceeding through all values of n, up to $n = 125$. This covers the entire procedure for the polynomial of order $\omega = 9$ in the variables u, v, w. A similar treatment was followed for $\omega = 10$, starting with $n = 125$ and iterating up to $n = 161$, and then for $\omega = 11$, iterating up to $n = 203$. A final refinement stage was calculated with double-precision arithmetic, in order to avoid possible rounding-off errors.

A detailed account of the results arising in the calculation would be too tedious to give here. Still, it seems relevant to provide some illustrative examples. For the iteration $n = 203$, for instance, Pekeris obtained for helium ($Z = 2$), a non-relativistic value of the energy parameter $\epsilon = 1.7040317568$ cm^{-1}. This was more accurate— Pekeris cared to stress—than those recently calculated with the help of electronic

---

[15] At the time, "double precision number" meant a quantity having twice as many digits as are normally carried in a specific computer. On the WEIZAC, normally, a number was represented by 40 binary digits. Today, IEEE standard defines "double precision" to include about 52 binary digits for the mantissa. The WEIZAC "double precision" supported more digits than IEEE standard. The WEIZAC fast memory contained 4096 words of 40 bits.

computers, among others by Toichiro Kinoshita (b. 1925) at Cornell, as well as by Hylleraas himself (Kinoshita 1957; Hylleraas and Midtdal 1956, 1958).

Additional important results concerned the ionization potential, which, as already explained, refers to the ratio of the least energy necessary for single ionization of the atom in the ground state. Pekeris indicated the need to consider the correction due to the Lamb Shift but warned that currently existing calculations were subject to uncertainties larger than those to which he had aimed at, namely 0.01 cm^{-1}. Herzberg had recently obtained an experimental result of $198{,}310.82 \pm 0.15$ cm^{-1} (Herzberg 1958), which was within the theoretical value now obtained by Pekeris, namely $I = 198{,}310.67 \pm 0.01$ cm^{-1}. In a sequel to his article published the following year, Pekeris went much further in his calculationss and obtained values for a determinant of order 1078. In this case, after considering the effect of the Lamb Shift, the value he obtained for the potential was $I = 198{,}310.674 \pm 0.025$ cm^{-1} (Pekeris 1959a, b). Beyond the extreme precision of this calculation in terms of the physical theory, the astounding size of the matrix (more than one million entries) was, in itself, an unprecedented achievement afforded by the availability of WEIZAC and the superb team that worked around it, particularly when considering that the RAM memory of the machine was merely 4096 words!

5.3   Aftermath

The amount of computing resources devoted to calculations related to the helium problem, as indicated in Table 5.2, was enormous.[16] From January 1958 to July 1959, for example, more than 40% of the WEIZAC processing time, in average, was dedicated to it.

And this was not yet the end of the project. In fact, over the next few years, Pekeris continued to perform even more exacting calculations that covered additional discrete states of helium (Pekeris 1959a), as well as of those of the lithium ion (Pekeris 1962) and several other interesting results as well.[17] By 1960, however, he had clearly realized that the computing power of WEIZAC would soon become a limitation, and that in order to make additional significant progress in the field, to cover all observed spectral lines of the helium, it would be necessary to rely on much faster machines. Construction of the next generation of computers was indeed a top priority in his program for DAM.[18]

Pekeris's methods and results continued to be cited for decades to come, in a surprisingly large number of research articles that praised them for their accuracy and

---

[16] The table was compiled from "Weizac Use Reports," (WIA 3–96-38 Pekeris: Computer Reports).

[17] Pekeris summarized his activities in this field in a report to the USA Air Force Office of Scientific Research, which sponsored his research from February 1961 to February 1963 (CPA).

[18] Scientific Activity Report 1960–1961, p. 5 (WIA).

Table 5.2 Computer time devoted to the helium project from January 1958 to July 1959

Month	Total computing hours	Helium project computing hours	%
Jul-59	661	214	32.38
May-59	555	254	45.77
Apr-59	661	282	42.66
Mar-59	676	300	44.38
Feb-59	593	305	51.43
Jan-59	609	307	50.41
Dec-58	616	319	51.79
Nov-58	608	318	52.30
Oct-58	663	351	52.94
Sep-58	483	233	48.24
Aug-58	603	204	33.83
Jul-58	596	319	53.52
Jun-58	589	278	47.20
May-58	641	234	36.51
Apr-58	611	143	23.40
Mar-58	546	235	43.04
Feb-58	553	133	24.05
Jan-58	461	240	52.06

originality (Qi-Hu and Zhong-Zhou 2012). Over the operational years of WEIZAC, namely between 1956 and 1963, a total of 14,000 articles were published in the *Physical Review*, and Pekeris's article is ranked in the 80th place, with 800 citations (Corry and Leviathan 2019, p. 70). The most remarkable testimony of the long-term impact of his work on the helium atom, however, is found in a huge volume published in 1995 by the American Institute of Physics to celebrate the centenary of that journal. This volume contains a selection of the most prominent and influential articles, about 1000 out of the quarter million, that appeared in the journal in its first hundred years of existence. Pekeris's article is featured in the chapter on atomic physics, introduced with the following description:

> In 1958 and 1959 Pekeris - using the Rayleigh-Ritz variational minimum principle and taking advantage of the vast improvements in computing power since the 1930s (though his computers were pitiful compared to today's, of course) - computed the energies of the ground and first excited states of atomic He to a heretofore unheard-of accuracy of about ten significant figures, including relativistic and other corrections; his calculations, wherein he diagonalized 1000×1000 matrices, illustrate the growing practice of expanding the sought-for solutions of the Schrödinger equation in function bases chosen less for their expected resemblance to the exact solutions than for their computational convenience. (Stroke 1995)

It is worth mentioning that with the continued increase in the computational power of electronic machines ever, more accurate approximations have been calculated for the ground state of the helium molecule, reaching up to 40-digits level

Fig. 5.4 Pekeris with Oppenheimer, Dedication of the Institute of Nuclear Science, May 20, 1958. Photographer Boris Karmi, Meitar Collection, The Pritzker Family National Photography Collection, The National Library of Israel

(Schwartz 2006). Some of these calculations rely on methods developed on the principles set down by Hylleraas and then further expanded by Pekeris (Zhang et al. 2019). Figure 5.4

References

Barrow-Green, J. 2008. The three-body problem. In *The Princeton Companion to Mathematics*, ed. T. Gowers, J. Barrow-Green, and I. Leader, 726–728. Princeton: Princeton University Pres.

Bartlett, J.H., J. Gibbons, and C.G. Dunn. 1935. The Normal Helium Atom. *Physical Review* 47 (9): 679–680.

Bloch, F. 1976. Heisenberg and the early days of quantum mechanics. *Physics Today* 29 (12): 23–27.

Burks, A.W., H.H. Goldstine, and J. von Neumann. 1946. *Preliminary Discussion of the Logical Design of an Electronic Computing Instrument*.

Ceruzzi, P. 2000. Nothing new since von Neumann: a historian looks at computer architecture, 1945–1995. In *The first computers: history and architectures*, 195–217. MIT Press.

Chandrasekhar, S., D. Elbert, and G. Herzberg. 1953. Shift of the 1 1S state of the Helium. *Physical Review* 91: 1172–1173.

Cohen, L., M. Scully, R. Scully. 2009. *Willis E. Lamb, Jr. A Biographical Memoir*. Washington, DC: National Academy of Sciences.

Coolidge, A.S., and H.M. James. 1937. On the convergence of the Hylleraas variational method. *Physical Review* 51 (10): 855–859.

Corry, L., and R. Leviathan. 2019. *WEIZAC: An Israeli Pioneering Adventure in Electronic Computing (1945–1963)*. Berlin: Springer.

Davis, P., and P. Rabinowitz. 1954. A multiple purpose orthonormalizing code and its uses. *Journal of the ACM (JACM)* 1 (4): 183–191.

Donnelly, J.R. 2011. Recollections of Chandra. *Physics Today,* 64(6): 8–10.

Erdelyi, A., W. Magnus, F. Oberhettinger, and F.G. Tricomi. 1953. *Higher Transcendental Functions*. New York: McGraw-Hill.

Eyring, H., J. Walter, and G.E. Kimball. 1944. *Quantum Chemistry*. New York: Wiley.

Fock, V.A. 1930. Näherungsmethode zur Lösung des quantenmechanischen Mehrkörperproblems. *Zeitschrift Der Physik* 61 (1): 126.

Forsythe, G.E. 1953. Solving linear algebraic equations can be interesting. *Bulletin of the American Mathematical Society* 59 (4): 299–329.

Frolov, A.M. 2006. Four-body perimetric coordinates. *Journal of Physics a: Mathematics and General* 39: 15421–15433.

Gavroglu, K., and A. Simões. 2011. *Neither Physics nor Chemistry: A History of Quantum Chemistry*. Massachusetts: MIT Press.

Goldberg, D. 1991. What every computer scientist should know about floating-point arithmetic. *ACM Computing Surveys (CSUR)* 23 (1): 5–48.

Hartree, D.R. 1928a. The wave mechanics of an atom with a non-Coulomb central field. Part I: Theory and methods. *Proceedings of the Cambridge Philosophical Society* 24: 89–110.

Hartree, D.R. 1928b. The wave mechanics of an atom with a non-Coulomb central field. Part II: Some results and discussion. *Proceedings of the Cambridge Philosophical Society* 24: 111–132.

Hartree, D.R. 1928c. The wave mechanics of an atom with a non-Coulomb central field. Part III: term values and intensities in series in optical spectra. *Proceedings of the Cambridge Philosophical Society,* 426–437.

Hartree, D.R., and W. Hartree. 1935. Self-consistent field, with exchange, for beryllium. *Proceedings of the Royal Society of London* 150: 9–33.

Hartree, D.R., and A. Porter. 1935. The construction and operation of a model differential analyser. *Memoirs and Proceedings of the Manchester Literary and Philosophical Society* 79: 51–73.

Hartree, D.R. 1929. The wave mechanics of an atom with a non-Coulomb central field. Part IV: further results relating to the optical spectrum. *Proceedings of the Cambridge Philosophical Society,* 310–315.

Hartree, D.R. 1947. *Calculating Machines: Recent and Prospective Developments and Their Impact on Mathematical Physics, Inaugural Lecture*. University Press.

Heisenberg, W. 1927. Mehrkörperprobleme und Resonanz in der Quanten-mechanik II. *Zeitschrift Für Physik* 41: 239–267.

Helgaker, T., and W. Klopper. 2000. Perspective on "Neue Berechnung der Energie des Heliums in Grundzustande", by Egil Hilleraas (1929). In *Theoretical Chemistry Accounts. New Century Issue,* ed. Cramer, C., and Trulhar D., 180–181. Berlin-Heidelberg-New York: Springer.

Herzberg, G. 1958. Ionization potentials and Lamb shifts of the ground states of 4He and 3He. *Proceedings of the Royal Society of London. Series A. Mathematical and Physical Sciences,* 248: 309–332.

Hund, F. 1927. Zur Deutung der Molekelspektren. *II. Zeitschrift Für Physik* 42: 93–120.

Hylleraas, E.A. 1928. Über den Grundzustand des Heliumatoms. *Zeitschrift Für Physik* 48 (7–8): 469–494.

Hylleraas, E.A. 1929. Neue Berechnung der Energie des Heliums in Grundzustande, sowie des tiefsten Terms von Orthohelium. *Zeitschrift Für Physik* 54: 347–366.

Hylleraas, E.A. 1930. Über den Grundterm der Zweielektronenprobleme von H−, He, Li+, Be++ usw. *Zeitschrift Für Physik* 65 (3): 209–225.

Hylleraas, E.A., and J. Midtdal. 1956. Ground-state energy of two-electron atoms. *Physical Review* 103: 829–830.

Hylleraas, E.A., and J. Midtdal. 1958. Ground-State energy of two-electron atoms. *Corrective Results. Physical Review* 109 (3): 1013–1014.

Hylleraas, E.A. 1964. The Schrödinger two-electron atomic problem. In *Advances in Quantum Chemistry,* ed. Löwdin, P.-O., Vol. 1, 1–34. Academic Press Inc.

James, H.M., and A.S. Coolidge. 1933. The ground state of the hydrogen molecule. *The Journal of Chemical Physics* 1 (12): 825–835.

Kellner, G.W. 1927. Die Ionisierungsspannung des Heliums nach der Schrödingerschen Theorie. *Zeitschrift Für Physik* 44 (1–2): 91–109.

Kinoshita, T. 1957. Ground State of the helium atom. *Physical Review* 105 (5): 1490–1502.

Kornerupi, P., and D.W. Matula. 2010. *Finite Precision Number Systems and Arithmetic.* Cambridge: Cambridge University Press.

Koutschan, C., and D. Zeilberger. 2011. The 1958 PEKERIS-ACCAD-WEIZAC Ground-breaking Collaboration that computed ground states of two-electron atoms (and its 2010 Redux). *The Mathematical Intelligencer* 33 (2): 52–57.

Kragh, H. 2012. *Niels Bohr and the Quantum Atom: The Bohr Model of Atomic Structure 1913–1925.* Oxford: Oxford University Press.

Lamb, W.E. 1951. Anomalous fine structure of hydrogen and singly. *Reports on Progress in Physics* 14: 19–63.

Lamb, W.E., and R.C. Retherford. 1947. Fine structure of the hydrogen atom by a microwave method. *Physical Review* 72 (3): 241–243.

Lyman, T. 1924. The spectrum of the helium in the extreme ultra-violet. *The Astrophysical Journal* 60 (1): 1–14.

Mawhin, J., and A. Ronveaux. 2010. Schrödinger and Dirac equations for the hydrogen atom, and Laguerre polynomials. *Archive for History of Exact Sciences* 64 (4): 429–460.

Mulliken, R.S., and C.C. Roothaan. 1959. Broken bottlenecks and the future of molecular quantum mechanics. *Proceedings of the National Academy of Sciences* 45 (3): 394–398.

Owens, L. 1986. Vannevar Bush and the differential analyzer: The text and context of an early computer. *Technology and Culture* 27 (1): 63–95.

Park, B.S. 2009. Between accuracy and manageability: Computational imperatives in quantum chemistry. *Historical Studies in the Natural Sciences* 39 (1): 32–62.

Pauling, L., and E.B. Wilson. 1935. *Introduction to Quantum Mechanics with Applications to Chemistry.* New York: Mcgraw-Hill.

Pekeris, C.L. 1946. Perturbation theory of the normal modes for an exponential M-curve in non-standard propagation of microwaves. *Journal of Applied Physics* 17: 678–684.

Pekeris, C.L. 1950. The zero-point energy of helium. *Physical Review* 79 (5): 884–885.

Pekeris, C.L. 1958. Ground state of two-electron atoms. *Physical Review* 112 (5). Retrieved from {https://link.aps.org/doi/10.1103/PhysRev.112.1649}

Pekeris, C.L. 1959a. 1 S 1 and 2 S 3 States of Helium. *Physical Review* 115 (5): 1216–1221.

Pekeris, C.L. 1959b. 1 S 1, 2 S 1, and 2 S 3 States of H- and of He. 126 (4): 1470–1476.

Pekeris, C.L. 1962. 1 S 1, 2 S 1, and 2 S 3 States of Li+. *Physical Review* 126 (1): 143–145.

Pekeris, C.L. 1972. Adventures in applied mathematics. *Quarterly of Applied Mathematics* 30 (1): 67–83.

Qi-Hu, L., and R. Zhong-Zhou. 2012. Three-Body Problem of H2+ Ion. *Communications in Theoretical Physics* 57: 214–216.

Sancho, P., and L. Plaja. 2008. Metastable superpositions of ortho- and para-Helium states. *Physics Letters A* 372 (34): 5560–5563.

Schwartz, C. 2006. Experiment and theory in computations of the He atom ground state. *International Journal of Modern Physics E* 15 (4): 877–888.

Schweber, S.S. 1990. The young John Clarke Slater and the development of quantum chemistry. *Historical Studies in the Physical and Biological Sciences* 20 (2): 339–406.

Slater, J.C. 1928. The normal state of Helium. *Physcial Review* 32: 349–350.

Stroke, H.H., ed. 1995. *The Physical Review. The First Hundred Years. A Selection of Seminal Papers and Commentaries.* Woodbury, NY: AIP Press.

Tanner, G., K. Richter, and J.-M. Rost. 2000. The theory of two-electron atoms: Between ground state and complete fragmentation. *Reviews of Modern Physics* 72 (2): 497–544.

Unsöld, A. 1927. Beiträge zur Quantenmechanik der Atome. *Annalen Der Physik* 82 (3): 355–393.

Wilkes, M.V. 1997. Arithmetic on the EDSAC. *IEEE Annals of the History of Computing* 19 (1): 13–15.

Wilson, S. 1997. Practical ab initio methods for molecular electronic structure studies. I. An overview. In *Problem Solving in Computational Molecular Science; Molecules in Different Environments,* eds. Wilson, S., and G.H. Diercksen 85–108.

Zhang, Y.Z., Y.C. Gao, L.G. Jiao, and Y.K. Ho. 2019. Linear dependence in Hylleraas configuration-interaction calculations of He atom. *Quantum Chemistry* 120 (7): 1–12.

Chapter 6
Additional Research with WEIZAC

Abstract WEIZAC was widely used in scientific computing projects led by scientists other than Pekeris. This includes work on Nuclear Magnetic Resonance, under the leadership of Shaul Meiboom and Shlomo Alexander, on ferromagnetic materials under the leadership of Shmuel Shtrikman and Ephraim H. Frei, and atomic spectroscopy under the leadership of Giulio Racah.

Keywords WEIZAC · Nuclear Magnetic Resonance · Ferromagnetic materials · Atomic spectroscopy · Shaul Meiboom · Shlomo Alexander · Shmuel Shtrikman · Ephraim H. Frei · Giulio Racah

In the preceding sections, we have described in detail the extent of Pekeris's intense use of WEIZAC and most of the important scientific outputs that he and his collaborators were able to achieve with it. Pekeris was the undisputed moving force behind the WEIZAC project and by all means the person who made the most intense use of the computing power of this machine. As already indicated in Table 3, from January 1958 to July 1959, for example, more than 40% of the WEIZAC processing time was dedicated to Pekeris's helium project. More broadly speaking, projects conducted at DAM consumed over that period more than 70% of the overall computing time.[1] Still, WEIZAC was also used by several government institutions in Israel, and—more closely related to the topic of this book—for scientific computing in projects led by other scientists. In this section, we provide brief outlines of some of the important, additional scientific projects that benefited at WIS from the computing power afforded by WEIZAC.[2]

---

[1] "Weizac Use Reports," (WIA 3-96-38 Pekeris: Computer Reports).

[2] For a comprehensive list of sixty-five scientific works completed with computations performed with WEIZAC, see Appendix B of Corry and Leviathan (2019).

L. Corry and R. Leviathan, *Chaim L. Pekeris and the Art of Applying Mathematics with WEIZAC, 1955–1963*, SpringerBriefs in History of Science and Technology, https://doi.org/10.1007/978-3-031-27125-0_6

6.1   Nuclear Magnetic Resonance

Nuclear magnetic resonance (NMR) phenomena were first detected experimentally in 1938 in a molecular beam. Its systematic study did not start until 1946, when two different groups of scientists, working independently in the US, discovered it for the case of bulk materials. One group was headed by Felix Bloch (1905–1983) at Stanford, in collaboration with William Hansen (1909–1949) and Martin Packard (1921–1989). The second group was from Harvard, and its leading members were Edward Mills Purcell (1912–1997), Henry Torrey (1911–1998), and Robert Pound (1919–2010). Bloch and Purcell were awarded the 1952 Nobel Prize in physics for "their development of new methods for nuclear magnetic precision measurements and discoveries in connection therewith" (Bloch and Purcell 2019). It took Bloch and Purcell a few months to recognize that they were observing two different aspects of the same phenomena, using two different experimental methods. Eventually, it was Bloch's approach that would become the standard one used in NMR spectrometers to this day (Becker 1993, 2000, pp. 2–5).[3]

Research activity using high-resolution NMR started at DAM as early as 1954 when a nuclear induction apparatus was built at WIS. In 1953, Bloch visited Rehovot, and in 1954, he joined the board of governors of WIS, becoming a long-time scientific advisor. Pekeris suggested to Bloch that Shaul Meiboom be sent to Stanford to study the subject under his guidance. Meiboom was born in Belgium and had studied physical engineering in the Netherlands. At the beginning of WWII, he moved to Palestine and later participated in the independence war. In the early 1950s, he joined DAM and participated in the geophysical project, while working on his Ph.D. that dealt with the theory of semi-conductors (germanium), under the formal guidance of Giulio Racah at the Hebrew University. Later, he headed the NMR group at DAM until 1958, when he permanently left WIS for the Bell Labs in the USA. In 1954, following Pekeris's request, Meiboom spent four months in Stanford, learning nuclear induction techniques. Upon return, the young Shlomo Alexander (1930–1998), who had just completed undergraduate studies at the Hebrew University, joined the department for Ph.D. studies. Part of his Ph.D. project with Meiboom was the building of a high-resolution spectrometer. The story of this important technology in Israel is similar to that of the WEIZAC: by mid-1955, a high-resolution nuclear induction apparatus was set up at WIS, which was a copy of Stanford's NMR spectrometer and one of the first high-resolution NMR spectrometers in the world. Electronics in Israel was then at its infancy, and electronics components were hard, or even impossible, to purchase, whereas the Stanford spectrometer was cutting-edge technology. The team used leftover components of the British army, so that the spectrometer screen, for instance, was originally the screen of a Radar (Luz 2007).

Among the first to use, the device was Israel Dostrovsky, who wanted to test hydrocarbons with NMR. The results obtained were complex spectra, which required calculations that could not be performed manually. Meiboom and Alexander learnt from Rabinowitz how to program WEIZAC, and the calculations were duly performed

---

[3] Bloch used the term "nuclear induction" and only later, in agreement with Purcell, they adopted the term "nuclear magnetic resonance (NMR)."

on the machine. The surprising results were reported in two important, highly cited papers (Alexander 1958, 1960).

In the summer of 1955, the notable chemist Ernest Grunwald (1923–2002) came from Florida to visit DAM. His first goal was to work with Dostrovsky and to use the isotope separation plant already operational at WIS. However, following a car accident in which (according to Zeev Luz's testimony) Dostrovsky broke a leg, being thus prevented from work at his lab, he decided to try the NMR, and he soon realized the many opportunities offered by it. This was the beginning of NMR chemical applications at WIS. Following Grunwald's advice, the group carried a set of experiments that led them to the discovery of methods for studying the kinetics of acid- and based-catalyzed proton transfer reaction in solutions, which were reported in three landmark papers: (Grunwald et al. 1957; Loewenstein and Meiboom 1957; Meiboom et al. 1958).[4]

The spectra were analyzed using parameters that were obtained from simulating on WEIZAC the exchange of broadened spectra. Members of the NMR lab solved the following problems with the aid of WEIZAC using the new core memory: (1) double quantum transition in nuclear magnetic, (2) calculation of exchange broadening in nuclear magnetic, and (3) calculation of complex spectra in nuclear magnetic resonance.[5]

Another visitor to the NMR lab in the summer of 1956 was Jerome Kaplan, from the Naval Research Laboratory in Washington DC. Following Meiboom's advice, he worked on developing a general theory for calculating dynamically broadened spectra (Luz 2007; Kaplan 1958). As noted in the scientific report 1956–1957, "a number of examples have been computed on the electronic computer."

David Gill (1929–1997) was another young graduate student who joined the group in 1958. Gill set out to measure NMR T_2-relaxation times in liquids, using the well-known Carr-Purcell pulse train scheme.[6] He found that the decay, which should be exponential, often exhibited peaks and other irregularities. Together with Meiboom, they developed an improved technique for measuring relaxation times. The technique was described in a paper that achieved more than 5000 citations (Meiboom and Gill 1958) and which is known under the name "Carr-Purcell-Meiboom-Gill" (CPMG) (Becker 2000, p. 235). In this case, calculations were not done with WEIZAC.

In the first years of the group's activity at WIS, it was part of DAM, working as a Nuclear Induction Laboratory. Then, in 1960, the Nuclear Magnetic Resonance Department was established as an independent unit. The coexistence of high-resolution NMR apparatus and an electronic computer in the same location during the late 1950s and early 1960s was a truly remarkable landmark that enabled innovative research in the new area of NMR at WIS and significantly contributed to the "dawn of NMR." For example, in order to interpret the observed spectra in terms of exchange rates, it was necessary to calculate theoretical spectra for many different cases. WEIZAC proved very valuable in reducing the necessary effort to a practicable

---

[4] We have gathered information about NMR in WIS by consulting (Luz 2007), as well as an interview by Leviathan with Luz, on Apr. 13, 2011.

[5] "Weizac Use Reports" 57-1956 (WIA).

[6] T_2-relaxation is the exponentially decay of the transverse component of the magnetization vector, in NMR.

amount.[7] Needless to stress here, NMR has become ever since an indispensable technique in a variety of scientific and technological fields, including organic chemistry,
structural biology, and anatomic imaging of human and animals. What the above
short account does stress, however, is the crucial role of WEIZAC in supporting the
early development of the field.

6.2   Magnetism and Ferromagnetic Materials

Computations performed with WEIZAC were at the basis of seminal work in the field
of ferromagnetics, which contributed to the understanding of the physical processes
underlying the functioning of magnetic storage devices. The main figures in this
project were Shmuel Shtrikman (1930–2003) and David Treves. Both of them grad-
uated from the Technion as engineers and joined WIS in 1954, at the newly established Department of Electronics. Shtrikman received a Ph.D. at WIS and went on
to become one of its leading researchers. Upon arrival at WIS, they joined Frei
(1912–2006) and his team, who were responsible for the construction of electronic
equipment.

In the course of their work with Frei, Shtrikman and Treves began their involvement with the issue of magnetism. In 1955, when WEIZAC was completed, they
had access to computing time, which they tried to use for solving the nonlinear
differential equations that describe the magnetic process of ferromagnetic particles.
In spite of the high-precision capabilities afforded by WEIZAC (40 bit), its double
precision subroutines, and the accuracy of its computations, the two did not succeed
in achieving results that came close enough to those predicted by existing theoretical models. Eventually, they also obtained an analytic solution. They published in a
seminal paper on micromagnetic analysis, authored also by Frei et al. (1957).

A methodology similar to the one developed at WIS was also introduced at roughly
the same time by William Fuller Brown (1904–1983), working in Minnesota (Brown
1957). Brown, who was to become a legendary figure in the fields of micromag-
netism and ferromagnetics, was incidentally asked to review the paper of the Rehovot
researchers. He immediately realized the importance of their results and insisted that
it be published as close as possible to his own one, which was already in press.
Shtrikman noted that, initially, Brown was doubtful about their results. In a later
paper that summarizes the work of the two groups, however, Brown wrote that the
calculations underlying them "are laborious, and (as far as I know) only Dr. Frei's
group and I have attempted them." Brown used an IBM 705 computer, which had
an accuracy of five decimal digits (which is about 16 binary digits) and hence was
lower than WEIZAC's 8 decimal digits and 16 double decimal digits.

According to Shtrikman's later testimony, their article was one of the most significant publications of the Department of Electronics in its first three decades, and he

---

[7] Scientific Activity Report, 1956–1957 (WIA).

stressed specifically that the results would not have been achieved without WEIZAC.[8] The article was widely publicized and cited by more than 700 scientific publications. It also led to a fruitful and ongoing collaboration with Brown. In 1962, Brown spent a year in Rehovot, and following this visit, he published one of his most cited articles:[9] "The Thermal Fluctuations of a Single-Domain Particle" (Brown 1963).

6.3 Atomic Spectroscopy

Another important research program conducted with the assistance of WEIZAC related to atomic spectroscopy. It was led by Giulio Racah (1909–1965), a prominent pioneer of physical research in Israel. Born in Florence, Italy, Racah immigrated to Palestine in 1939, after working as Fermi's assistant in Rome and Pauli's in Zürich, and after teaching at the Universities of Florence and Pisa. In 1949, he was appointed to the chair of theoretical physics at the Hebrew University, the institution that he also led as rector in the years 1961–1965. He educated generations of physicists and contributed to put Israel on the map of world physics (Unna 2000; Zeldes 2009). Racah developed powerful computational methods to address theoretical questions in atomic and nuclear physics and—together with his students and collaborators and the crucial support of WEIZAC—performed many crucial calculations in a discipline that was then at the center of attention in physics worldwide.

In order to understand the extent of Racah's influence on physics in Israel, and the way in which WEIZAC enters this picture, it is important to describe in the first place the institutional context within which he operated. Racah supervised in Jerusalem 58 MSc theses and 15 Ph.D. theses. Of these, 46 and 8, respectively, comprised theoretical analyzes of experimental atomic spectra that were performed using the methods that Racah had developed (Zeldes 2009, p. 208). In addition, Racah was a member of the Israel Atomic Energy Commission in the years 1952–1955 and a member of Israel delegations to the first three Geneva conferences (1955, 1958, 1964) on the peaceful uses of atomic energy.

Most importantly, Racah was a member of the scientific committee of the Ministry of Defense in 1948–1952. Following his initiative, Ben-Gurion in 1949 authorized HEMED, the Science Corps of the IDF (Israel Defense Forces), to fund the postgraduate work in nuclear physics of six promising physics graduate students who had just served in the war. Racah made sure that they would go to the world's leading institutions and get training with the leading physicists of their time, mostly Nobel Prize awardees. Thus, Amos de Shalit (1926–1969) went to the ETH, Zürich, to study with Paul Scherrer (1890–1969). Igal Talmi (b. 1925) became a student of Wolfgang Pauli at the same institution. Uri Haber-Schaim (1926–2020) went to pursue his studies

---

[8] "WEIZAC and Golem pioneers." Video interviews partially supported by the American Committee for the Weizmann Institute of Science—private copy (Also available at https://ethw.org/Archives: The_Computer_Pioneers,_Weizmann_Institute_segment_3).

[9] More then 4700, according to Google Scholar.

under Enrico Fermi in Chicago. Gideon Yekutieli (1926–1999) worked on experimental physics in Bristol with Cecil Frank Powell (1903–1969). Gvirol Goldring (b. 1926) worked on experimental nuclear physics at Imperial College, London, under Samuel Devons (1914–2006), and Israel Pelah (Pelchovitch) (1923–1982) studied in Amsterdam, Holland (Cohen 1999, p. 26).

In 1953, after WIS had opened departments of Applied Mathematics, Biophysics, Experimental Biology, Isotope Research, Optics Organic Chemistry, and Polymer Research, it was felt that the next step should involve a Department of Physics. The topic was discussed in November 1953, at a meeting of the Scientific Committee, which counted among the participants, as guests from Copenhagen, Niels Bohr, and Aage Bohr (his physicist son, who in 1975 would share the Nobel Prize). Biophysicist Ephraim Katchalsky (Katzir) (1916–2009) served as chair of the committee, and in emphasizing the need to include Nuclear Physics in the planned department, he referred to Racah's activities, in the following terms:

> Prof. Racah had developed at the Hebrew University a good school of Theoretical Physics, and it would no doubt be to the advantage of all concerned if we could also have a Department of Experimental Physics with which theoreticians could collaborate to the advantage of all concerned.[10]

Niels Bohr indicated two fields that in his opinion should receive special attention in any institution involved in nuclear research: atomic piles and the spectroscopy of energy of emitted particles. Concerning the crucial issue of qualified faculty members, Katchalsky mentioned, "a few physicists abroad studying pile techniques and problems, and these would later have the responsibility for training others. Some of these physicists would be working incidentally on cosmic radiation and spectroscopy." And as a fundamental issue to be handled Bohr also insisted on the need to count with an adequate calculating machine, "they in Copenhagen—he said—relied on the machine in Lund."[11]

And indeed, as soon as the said physicists returned to Israel, they were hired as the team that in 1954 established at WIS what would become a world-class center of nuclear physics. Moreover, given their personal attachment to Racah and his research program, the Rehovot group and the department of theoretical physics in Jerusalem kept regular working contact, at the center of which was the joint weekly seminar at WIS.

A curious related point arising in the report concerns Bohr's opinion as to the possible military uses of atomic energy, which is described in the following terms:

> The idea of Israel engaging in military atomic energy work was of course hopeless. In money alone, such a project would cost Israel's total national income for years, and this was quite apart from the other difficulties of man-power, space, etc. He sincerely hoped that by the time Israel could possibly achieve anything in this field, the world would have settled many of its military and political problems. As he had already said, there did exist good fields of research at lower energies, though these could still be costly. For such work, Rehovot might well be a suitable place.

---

[10] Minutes of the meeting of the Scientific Committee held on Nov. 9, 1953 (CPA).

[11] Minutes of the meeting of the Scientific Committee held on Nov. 9, 1953 (CPA); Scientific Activity Report 1954 (WIA).

If the report reflected faithfully what Bohr actually said to the members of the committee, then one wonders what some of them thought in silence, as they were directly involved in Israel's then secret project of atomic weapons (e.g., Katchalsky himself, his brother Aharon, and Dostrovsky).

Now, theoretical spectroscopy was one of Racah's main fields of activity, where he specialized in studying properties of atoms via their spectra by combining theoretical, computational, and experimental methods. As already mentioned in relation to Pekeris's work on the helium atom, Niels Bohr's model of the hydrogen atom, introduced in 1913, clarified the correlation between the presence of spectral lines and the passage of the electron from separate, discrete energy levels within the atom. This field of study gained enormous impulse following the creation in the early 1930s of quantum mechanics and the gradual development of ever more sophisticated and accurate measurement techniques for spectra. Based on the study of the interaction between matter and electromagnetic radiation, atomic spectroscopy focused on representing the interplay between many-electron atoms and radiative energy as a function of wavelength or frequency. By doing so, it afforded from very early on a flurry of results of great interest, based on the very high level of experimental precision that characterized it.

Nuclear spectroscopy developed somewhat later by focusing on the energy levels of nuclei and of the transitions between them. In both the atomic and the nuclear cases, energy levels take only discrete values, such values attaining much higher figures in the nuclear case than in the atomic one. By 1955, nuclear spectroscopy had not achieved the levels of precision already known in the atomic case, and the internal mechanics of nuclei were much less understood than what the current knowledge about atoms and their structure afforded. Quantum mechanics was considered at the time to be a well-established field of knowledge, whereas the study of the laws governing the forces between the constituents of the nucleus was at a rather incipient stage. Still, the technical and theoretical developments recently associated with nuclear spectroscopy were considerable (Pryce 1957).

In 1935, Condon published in collaboration with George H. Shortley (1910–1980) a textbook that became the standard comprehensive work on the theory of atomic spectra for decades to come (Condon and Shortley 1935). Based on an exposition of theoretical quantum mechanics along the lines formulated by Paul Dirac in his landmark book of 1930 (Dirac 1930), it presented a unified, logical deduction of the structure of the spectra of atoms. At the time of its publication, it was possible to give a reasonably complete account of the subject, and this text did so in a most accurate manner while making little concession to the reader. Among other things, it presented a classical version of a numerical method previously developed by Slater that allowed for approximated calculations of energy levels but was limited to relatively simple cases of two-electron configurations. At the same time, however, the book fell short of taking full advantage of the important mathematical insights afforded at the time by modern algebraic theories, and particularly group theory, as the appropriate way to express the fundamental symmetries characteristic of the physical processes involved (Judd 1975).

This was the starting point of Racah's groundbreaking contribution, which was initially published in 1942–1949 in the *Physical Review* in four classical papers known as "Theory of Complex Spectra I–IV" (Racah 1942–1949). Racah came up with analytic solutions to what Slater's method could solve only numerically and developed powerful mathematical methods for calculating atomic spectra that could yield concrete results, applicable in a wide range of more complex situations. These methods relied on the use of the recently introduced algebraic tools, in particular those relying on the theory of continuous groups (or Lie groups) to calculate highly complicated spectra. While Racah's interest focused mainly on the physical questions related to the interpretation of atomic spectra, his contribution to the development of group-theoretical methods had great inherent mathematical interest. His research had a significant impact on the general theory on Lie groups, the classification of the (complex) semisimple algebras and the representation theory associated with the latter. His Complex Spectra papers laid the ground for what is known today as "Racah Algebras."

The mathematical core of Racah's papers introduced new, powerful techniques for computing energy levels in terms of scalar products of tensor operators, and they considerably simplified computations of matrix elements. In earlier approaches, the computations of more complex spectra rapidly became too cumbersome, and indeed prohibitive, so that they could not be carried out in practice well. Racah also introduced a new quantum number, *seniority v*, which became crucial for solving the designation problem of energy levels in configurations having several equivalent electrons (Racah 1958a). These notions were applied to atomic spectra in the first place, but from the mid-1950s onwards, they turned out to be essential in nuclear physics as well.

For a short while after 1949, Racah's results remained unnoticed in atomic spectroscopy, but their importance in nuclear physics was rapidly recognized and applied to specific situations. The four papers revolutionized the field, and they constitute an essential research tool in nuclear physics and the physics of elementary particles to this day.

At the time, however, several computational challenges still lay ahead that could only be met efficiently as the first big electronic computers became available. Efficient diagonalization of matrices of very high orders was foremost among these challenges. Diagonalization procedures that had been hitherto avoided for reasons of computational difficulty now became possible. New results could be obtained that would allow for comparing, to an unprecedented extent and accuracy, theoretical predictions with the experimental results. Racah was quick to understand the newly created situation and eager to seize the opportunity.

Before entering the era of digital electronic automatic computers, most calculations appearing in Racah's papers were performed with the help of mechanical calculators, such as the Odhner,[12] or of electro-mechanic ones, such as the Madas 20ASZ.[13] These allowed for the study of relatively simple spectra, with energy matrices of limited size. For matrices of order 23 or 25, these machines could not properly deliver the desired results within reasonable time periods (Racah 1955).

---

[12] http://www.vintagecalculators.com/html/odhner.html.

[13] http://www.vintagecalculators.com/html/madas_20azs.html.

In 1954, Racah initiated a collaboration with William Meggers (1888–1966) who at the time was a leading figure in the field of spectroscopy.[14] Indeed, Meggers was head of the spectroscopy section at the Atomic and Radiation Physics Division of the National Bureau of Standards (NBS) in Washington, DC, where a powerful digital electronic computer, Standards Eastern Automatic Computer (SEAC), had been operational since May 1950.[15] In the framework of this collaboration, it became possible for the first time to diagonalize increasingly large matrices, of order up to 38 × 38, on behalf of calculations related with Racah's work. Interestingly, Rabinowitz also participated in this effort, and in the publications, he already indicated WIS as his institutional association (Trees et al. 1955; Zeldes 2009, p. 10).

A general feature of articles published in this period, which provide early examples of the use of electronic computers for solving open scientific problems, is that they typically provide interesting historical insights about the uneasy passage to the new digital era but also of the incredible ability for quick adaptation noticeable in the work of some scientists. Racah provides a most prominent example of this. In a footnote to one of the articles related to the collaboration between his team and the NBS, for example, we find the following interesting testimony (Trees et al. 1955, p. 337 footnote 15):

> The time required to set up the code was naturally far in excess of the time needed to do the work with hand computers. The expenditure of this time is justified only because it is expected that the same coding will be used subsequently in many similar problems.

Over the following two years, Racah developed additional programs to diagonalize matrices. The results obtained on the basis of this diagonalization were successfully compared with experimental data. One of the important outcomes of the introduction of computers in this field concerned problem choice. If up to this point, the agenda of research was established on the basis of such problems that could be solved with the available machines, which were less powerful, now the focus moved to those problems which posed real, inherent scientific challenges.

Upon completion of WEIZAC, Racah becomes an assiduous user, in collaboration with his students, who had already become part of the physics department at WIS. Racah's first computer program was sent by mail from Jerusalem, and after being typed at DAM, it ran flawlessly on the first attempt.[16] Coming once a week to Rehovot, Racah conducted much of the programming work himself and was never deterred from correcting the program "in real time," in cooperation with Riesel (Shadmi 1965).

Working close to home, rather than overseas, allowed Racah and his team to continually and considerably improve the methods and thus achieve increasingly

---

[14] https://www.nist.gov/director/nbsnist-culture-excellence/william-F-meggers-dean-american-spectroscopists.

[15] See (Kirsch 1998). At roughly the same time, also the Standards Western Automatic Computer (SWAC) became operational in the West Coast (Corry 2008, 40–48). They were the fastest operational computing machines before the IAS computer became operational one year later. Both machines played an important role in the widespread adoption of electronic computing in the USA, both in science and in administration.

[16] Aviezri Fraenkel, interview by Leviathan, Feb. 8, 2010.

accurate results in light of the experience gained over two years of intense work. The team ran three main kinds of programs on WEIZAC. One was used for the diagonalization itself. A second one was used for comparing the results of the diagonalization with the experimental data and for improving the values of the so-called "interaction parameters." This was done with the help of the method of least squares. There was also a program for calculating line strengths (Racah 1958b).

Racah was very sensitive about questions of precise programming techniques and adequate use of resources. According to the testimony of Meir Weinstein, one of the early operators of WEIZAC from very early on Racah adopted the innovative strategy of breaking the execution of programs into several separate runs, while keeping intermediate results. This was meant to prevent the need to rerun the program in the event of a hardware malfunction.[17] In Racah's own words (Racah 1958b, pp. 1–2):

> The main lesson derived from our working experience was that each program should be able to perform, in addition to its main tasks, a series of auxiliary operations for simplifying the form in which the data may be fed, transforming the input data to the form which suits best the specific problem, elaborating the results in order to obtain all the information which may be needed, and, last but not least, checking the data and the results against possible errors.

Racah also suggested that the best way to use memory space ("which is fairly big, but not unlimited"), would be to assign this resource in a dynamic, rather than pre-determined, manner. That is, it should be possible to allow the operator to assign memory space to the different tasks according to the dimensions of the specific problem currently run.

Pekeris was proud to mention Racah's program among the important uses of WEIZAC.[18] He described Racah's approach as a semi-empirical one, based on a theory in which one assumes in the zeroth approximation that the electrons move in a central field produced by the nucleus and by the mean action of the other electrons, while treating the residual interaction between the electrons as residual. The related calculation is performed in first approximation, and then a semi-empirical parameter takes into account most of the contributions of the higher approximations.

The largest problems solved by Racah and his collaborators with the help of WEIZAC involved systems of 650 equations with 36 unknowns, in which powerful methods for diagonalization were implemented. They were part of an ongoing survey of various kinds of spectra, from which it became increasingly clear that the parameters that measure the interaction between the different electrons of an atom can be evaluated by simple interpolation formulas and that unknown energy levels can be predicted with very good approximation rates (Racah 1958; Racah and Shadmi 1959; Racah and Spector 1960).

After WEIZAC was shut down in 1963, the Department of Physics of the Hebrew University continued to pursue its computational tasks with the help of a PHILCO 2000 computer purchased by the Ministry of Defense. Later, in 1965, shortly before Racah's untimely death, they switched to an IBM 7040 that was acquired by their own institution.

---

[17] Weinstein, interview by Leviathan, Jan. 16, 2012.

[18] In his Report of 1960 (CPA).

These were, then, the main, but not the only, projects carried out by scientists other than Pekeris himself, and where calculations with WEIZAC played a fundamental role. Each of them deserves to be discussed in greater detail, but this is left for a future opportunity. We should also mention, at least in passing, the problem of the interpretation of X-ray diffraction patterns used for determining the structure of crystals, to the mathematical side of which Joseph Gillis made important contributions (Gillis 1958). A completely different project was the Responsa Project, whose main driving force was Aviezri Fraenkel (b. 1929), one of the first WEIZAC designers. The Responsa Literature was a corpus of Jewish case law that played an important role in the life of communities throughout the centuries, particularly in Europe. It comprised the accumulated collection of answers and rulings given by prominent rabbis in reply to written questions on halakhic issues sent to them by individuals in need of counsel. Fraenkel came up with the idea that the best way to make these rulings readily available to Jews all over the world looking for answers to their queries was with the help of a computer-assisted system. The system, first introduced in 1968, underwent various upgrading stages over the following decades (Corry and Leviathan 2019, p. 51).

References

Alexander, S. 1958. Relative signs of spin-spin interaction in nuclear magnetic resonance. *The Journal of Chemical Physics* 28 (2): 358–359.

Alexander, S. 1960. Relative signs of spin-spin interaction in nuclear magnetic resonance. II. *The Journal of Chemical Physics* 32 (6): 1700–1705.

Becker, E.D. 1993. A brief history of nuclear magnetic resonance. *Analytical Chemistry* 65 (6): 295A-302A.

Becker, E.D. 2000. *High Resolution NMR: Theory and Chemical Applications*, 3rd ed. New York: Academic Press.

Bloch, F., and E.M. Purcell. 2019. *Felix Bloch—Biographical. NobelPrize.org. Nobel Media AB 2019*. Retrieved from https://www.nobelprize.org/prizes/physics/1952/bloch/biographical/

Brown, W.F. 1957. Criterion for uniform micromagnetization. *Physical Review* 105 (5): 1479–1482.

Brown, W.F. 1963. Thermal functuations of a single-domain particle. *Physical Review* 130: 1677–1686.

Cohen, A. 1999. *Israel and the Bomb*. New York: Columbia University Press.

Condon, E.U., and G.H. Shortley. 1935. *The Theory of Atomic Spectra*. Cambridge: Cambridge University Press.

Corry, L. 2008. Fermat meets SWAC: Vandiver, the Lehmers, computers, and number theory. *IEEE Annals of the History of Computing* 30 (1): 38–49.

Corry, L., and R. Leviathan. 2019. *WEIZAC: An Israeli Pioneering Adventure in Electronic Computing (1945–1963)*. Berlin: Springer.

Dirac, P.A. 1930. *The Principles of Quantum Mechanics*. Oxford: Clarendon Press.

Frei, E.H., S. Shtrikman, and D. Treves. 1957. Critical size and nycleation field of ideal ferromagnetic particles. *Physical Review* 106 (3): 446–455.

Gillis, J. 1958. An application of electronic computing to X-ray crystallography. *Acta Crystallographica* 11 (12): 833–834.

Grunwald, E., A. Loewenstein, and S. Meiboom. 1957. Rates and mechanisms of protolysis of methylammonium Ion in aqueous solution studied by proton magnetic resonance. *The Journal of Chemical Physics* 27 (3): 630–640.

Judd, B.R. 1975. Perspectives on "The theory of atomic spectra." In *Atomic Physics 4*, 13–17. New York: Springer.

Kaplan, J. 1958. Exchange broadening in nuclear magnetic resonance. *The Journal of Chemical Physics* 28 (2): 278–282.

Kirsch, R.A. 1998. SEAC and the start of image processing at the National Bureau of Standards. *IEEE Annals of the History of Computing* 20 (2): 7–13.

Loewenstein, A., and S. Meiboom. 1957. Rates and mechanisms of protolysis of Di-and Trimethy-lammonium ions studied by proton magnetic resonance. *The Journal of Chemical Physics* 27 (5): 1067–1071.

Luz, Z. 2007. The early days of NMR in Israel. In *Encyclopedia of Magnetic Resonance*. Wiley Online Library.

Meiboom, S., and D. Gill. 1958. Modified spin-echo method for measuring nuclear relaxation times. *Review of Scientific Instruments* 29: 688–691.

Meiboom, S., A. Loewenstein, and S. Alexander. 1958. Study of the protolysis kinetics of ammonium ion in aqueous solution by proton magnetic resonance technique. *The Journal of Chemical Physics* 29 (4): 969–970.

Pryce, M. 1957. Recent advances in nuclear spectroscopy. *Rendiconti Del Seminario Matematico e Fisico Di Milano* 27 (1): 3–16.

Racah, G. 1942–1949. Theory of complex spectra , I–IV. *Physical Review* 61, 62, 63, 76: 186–197; 438–462; 367–382; 1352–1365.

Racah, G. 1955. The present state and problems of the theory of atomic spectra. In *Proceedings Rydberg Center conference on atomic spectroscopy*, ed. E. Bengt, Vols. Kgl. Fysiogr. Sällsk. Handl. 65, Nr. 21, pp. 31–42.

Racah, G. 1958a. The seniority quantum number and its applications to nuclear spectroscopy. In *Proceedings of the Rehovoth conference on nuclear structure*, ed. H.J. Lipkin, pp. 155–160. Amsterdam: North-Holland.

Racah, G. 1958b. The use of the WEIZAC in theoretical spectroscopy. *Bulletin of the Research Council of Israel* 8F: 1.

Racah, G., and N. Spector. 1960. The configurations 3d" 4p in the second spectra of the iron group. *Bulletin: Mathematics and Physics* 9, 75.

Racah, G., and Y. Shadmi. 1959. The configurations (3d+ 4s) n in the second spectra of the iron group. *Bulletin Research Council of Israel F* 8: 15–46.

Shadmi, Y. 1965. The physicist (יאקיזיפה). In *On Professor Yoel Racah*, 15–21. Jerusalem: Magnes [In Hebrew].

Trees, R.E., W.F. Cahill, and P. Rabinowitz. 1955. Computation of atomic energy levels: Spectrum of singly-ionized Tantalum (Ta II). *Journal of Research of the National Bureau of Standards* 55 (6): 335–341.

Unna, I. 2000. The genesis of physics at the Hebrew University of Jerusalem. *Physics in Perspective* 2: 336–380.

Zeldes, N. 2009. Giulio Racah and theoretical physics in Jerusalem. *Archive for History of Exact Sciences* 63: 289–323.

Chapter 7
Mathematics at WIS After WEIZAC, Applied, and Pure

Abstract WEIZAC was shut down in 1963. By that time Pekeris's success was already internationally acknowledged. Efforts to construct the new generation of computers had become a top priority at the Weizmann Institute as it was felt that the computing power WEIZAC had started to become a limitation. At the same time, plans for creating a department of pure mathematics started to be debated. Pekeris opposed the move, but the department of pure mathematics became fully operative in 1971.

Keywords WEIZAC · GOLEM A-B · Chaim L. Pekeris · Amir Pnueli

When WEIZAC was shut down in 1963, Pekeris had every thinkable reason to believe that the plans he had devised back in 1948 when joining WIS had been crowned with full success. He could be proud for having established a center for excellence in applied mathematics in Israel at the center of which would stand an electronic computer. In fact, by 1962 Pekeris's success was already internationally acknowledged. Interesting evidence for this appears in the following statement of 1962 by Chandrasekhar:

> I think it can be fairly said that the record of what has been accomplished at the Weizmann Institute with the WEIZAC under the leadership of Professor C. L. Pekeris is unequalled in the world. The uniqueness of this accomplishment derives not so much from the high quality or the large quantity of work that has been done as from the fact that Pekeris and his associates have for the first time used an electronic computer for the solution of problems which one could not literally have dreamt of solving before.[1]

In addition to its undisputable successes, his department also posed an alternative to the neo-humanistic ethos of research in pure mathematics that had already been built with great success in Jerusalem. As already indicated, Pekeris saw himself as belonging to a tradition in science that followed on the footsteps of Lord Rayleigh, where physical research and applied mathematics went hand in hand. He believed he was now at the high point of a process meant to overcome the existing gap across the two disciplines as previously pursued in Israel. In his survey of research done at DAM

---

[1] Chandrasekhar, "A memorandum on the Application from the Weizmann Institute of Science for a High-Speed Electronic Computer," 1962 (CPA). Cited in (Corry and Leviathan 2019, p. 72).

© The Author(s), under exclusive license to Springer Nature Switzerland AG 2023   111
L. Corry and R. Leviathan, *Chaim L. Pekeris and the Art of Applying Mathematics with WEIZAC, 1955–1963*, SpringerBriefs in History of Science and Technology,
https://doi.org/10.1007/978-3-031-27125-0_7

during its first 7 years, entitled "Fundamental Research in Applied Mathematics," he explicitly addressed that issue when he wrote:

> The principal field of interest in our fundamental research program might be termed classical physics. We are faced with the situation that, on the one hand, mathematical training in Israel is almost exclusively restricted to pure mathematics, and that, on the other hand, the trend among physicists, both young and old, is to crowd into the physics of nuclear physics. We have therefore taken upon ourselves the task of fostering research, as well as advance training, in the broad field of classical physics. Current topics under investigation are: electromagnetic wave propagation, theoretical seismology, tides in the atmosphere, oceans and earth, hydrostatic stability of superposed fluids, problems of random walk and Brownian motion, and the phase problem in X-ray crystallography.[2]

As already indicated at the end of Chap. 5 above, by 1960 efforts to construct the new generation of computers had become a top priority at DAM, as it was felt that the computing power WEIZAC had started to become a limitation. In fact, the plans for building a new computer at WIS, the GOLEM, had already begun two years after WEIZAC had become operational,[3] and the design was started with the arrival of Smil Ruhman (b. 1925) acting as the chief engineer of the project. While those plans were underway, it was decided in early 1961 at WIS to turn to the government in order to use computing time of the IDF's TRANSAC,[4] a Philco computer, which was the first electronic machine acquired by the Israeli government, and the first one in Israel outside WIS (Shahar 2002). A few months before the arrival in Israel of TRANSAC, Pekeris started negotiations with the first commander of MAMRAM (the IDF Computer Unit), Mordechai Kikayon (1915–1993). Pekeris referred to him as "the person in charge of the Government computer." He asked to be assigned future TRANSAC hours in exchange for WEIZAC hours "at the rate of four to one." According to Pekeris, TRANSAC was ten times faster than WEIZAC.[5] Pekeris involved Shimon Peres (1923–2016), then the undersecretary of defense, as well as Abba Eban (1915–2002), then the president of WIS, in his efforts to secure future computer hours.

Using the full power of his influence at WIS, Pekeris went on to reach a formal agreement meant to create a new bank account in the USA, fully dedicated to this project, of which the only, joint authorized signatures were his and that of Weisgal, Chairman of the Executive Council, and the strongman of WIS.[6] It was agreed that "the money in this account will not be used as collateral for loans or for any other purpose that might endanger or delay the instant availability of the cash when needed." Even more remarkable is the fact that the money was to be obtained from contracts signed with ARPA and NFS, motivated by the desire of these two powerful

---

[2] Pekeris, "Survey of research done by department of applied mathematics at Weizmann Institute of Science during the period of 1949–1956" (CPA).

[3] Zwi Riesel to Pekeris, Long range plans for the Computer Group, June 16, 1957 (CPA).

[4] Scientific Activity Report 1960–1961, p. 7 (WIA).

[5] Pekeris to Remez (Administrative Director of WIS), Aug. 30, 1959 (CPA).

[6] The centrality of Weisgal's support for the success of the WEIZAC is extensively explained throughout the chapters of (Corry and Leviathan 2019).

USA institutions, "to extend help for the construction of computer NIC at the Weizmann Institute." New Illinois Computer (NIC) was a computer then under construction at the University of Illinois that was to be copied at WIS.[7] The agreement further established that Pekeris and Eban would use their influence at the government level[8] to ensure that "their income from computing on TRANSAC accruing from the ARPA and NSF contracts be returned to the Weizmann Institute for the purpose of purchasing parts for NIC." In turn, WIS would repay the government "in computing time on NIC at an exchange rate of NIC versus TRANSAC time" in terms to be discussed in the future. The contract with ARPA and NSF concerned research projects such as the determination of oceanic tides, problems in theoretical seismology, the three-body problem in atomic physics, and others. Over the years 1961–1963, a few hundred TRANSAC hours were used on behalf of these projects.[9]

When the operation of WEIZAC was terminated in 1963, a CDC-1604 computer had been donated to WIS to be used as an intermediate solution until the construction of a new computer would be completed. The dollars accumulated in the special fund were used for purchasing parts for the GOLEM. Although its logical structure was based on that of NIC, the GOLEM's design included new kinds of circuitry aimed at greater economy and reliability, as well as compactness of packaging. It incorporated new transistors, which had become available in the meantime.[10] GOLEM was activated in 1965, and due to the lower costs of its components, a second machine, which was an exact copy of it, was built and activated in 1966. The design of yet another computer, GOLEM B, was initiated right away, but the project encountered unanticipated difficulties, and the design was completed only in 1974, by which time an IBM computer-based computation center was already in use at WIS (Corry and Leviathan 2019, pp. 66–67).

Also, in the later phase of his career, Pekeris continued to exploit the full capacities of his expertise with electronic computers, while paying particular attention to the theory of ocean tides. In the already mentioned "Pekeris Memorial Lecture" of 1995, Lighthill stated that Pekeris' contributions in this area were of "especial excellence." A joint 1969 paper with Accad on the "Solution of Laplace's equations for the M_2 tide in the world oceans" (Pekeris and Accad 1969)—Lighthill stressed in particular—was of "revolutionary importance" (Lighthill 1995, p. 12). In fact, back in 1775, Laplace had initiated the study of tides considered from the point of view of a moving mass of water, but the hydrodynamic equations that describe the motion of tidal waves across the ocean remained unsolved, in spite of successive attempts to obtain approximate predictive solutions. Pekeris and Accad were the first to use

---

[7] Pekeris to Weisgal ("Summary of agreement reached on November 9, 1960 concerning establishment of a trust fund for the new electronic computer NIC which is to be built by the department of applied mathematics of Weizmann Institute"), Nov. 14, 1960.

[8] Eban was a highly influential figure in Ben Gurion's Mapai Labor Party. He had been Ambassador to the USA and would later serve as Member of the Knesset and Foreign Minister, among other important positions.

[9] From Treasury and Budget Section to Pekeris, Re: Sinking Fund, June 7, 1962. Work Plan for MAMRAM for the years 1962/1963 Nov. 20, 1961 (IDF Archive 410-207-1970, pp. 158–159).

[10] Scientific Activity Report 1962–1963 p. 17. Ruhman, interview by R. Leviathan, Nov. 13, 2013.

Laplace tides equations for a realistic ocean model. Lighthill, in his *The First Chaim Leib Pekeris Memorial Lecture* (Lighthill 1995), considers this as "a very special achievement of Pekeris and Accad (1969)."

It is also pertinent to indicate that one of Pekeris's students at the time, who participated in the work on tides, was Amir Pnueli (1941–2009), a recipient of the Turing Award in 1996. His PhD dissertation was on the topic of "Solution of tidal problems in simple basins," and it was followed by two important publications (Pnueli and Pekeris 1968a, b).

In parallel to the new projects, Pekeris engaged in a very vivid, long-term, and intense correspondence with experimentalists working around the world, particularly in matters related to the helium atom. He was proud to declare the "unique situation" that his work had achieved, whereby "for the first time the experimentalists were lagging behind the theoreticians in accuracy" (Pekeris 1972, p. 78). In a public lecture of 1972, Pekeris declared that the field of atomic spectroscopy remained "the burden of the applied mathematician for the future," and he was able to formulate in detail the basic problems that, even in two-electron spectroscopy, remained then unsolved. The Laguerre functions adopted in Eq. (5.4), for instance, raised the question whether there could be a more adequate kind of expansion, more intrinsically related to Eqs. (5.2) and (5.3) of Chap. 5, and leading to more accurate results. Pekeris expressed the hope that additional effort would soon be devoted to elucidating these and additional, related open problems.

Pekeris remained in his position as the director of DAM until his formal retirement in 1973, but he continued to be involved in the activities of the department as well as in active research as "Distinguished Institute Professor." In fact, beginning in 1976 he also published several papers in what was for him a new field of research, relativity theory. All of his last published papers were in this field, including the last one in 1993 (Pekeris and Frankowski 1993), the proofs of which reached him just few days before his death at the age of 85 (Artstein 1995).

Pekeris was a member of many prominent scientific associations throughout the world, including, among others, the National Academy of Science in the USA, the Royal Astronomical Society, and the Academia dei Lincei in Rome. He received important awards in Israel (such as the Rothschild Prize in Mathematics in 1966 and the Israel Prize in Physics in 1980), as well as abroad (such as the Vetlesen Prize for Earth Sciences in 1974 and the gold medal of the Royal Astronomical Society in 1980).

But in spite of all his impressive achievements and accolades, Pekeris was well aware of the kind of reservations with which his projects met among mainstream mathematicians. Back in 1956, he had addressed this issue head-on, in a survey on the activities of DAM:

> This brings into focus a concern which is uppermost in the minds of those applied mathe-
> maticians who live by the lustre of their analytical prowess, namely, whether the electronic
> computer is not destined to mechanize mathematical analysis, and thereby obliterate it. My
> limited experience with our mechanical brains leads me to the belief that the contrary is true.
> In the case of the earthquake problems, for example, the computer produced certain results
> which I had not foreseen. Before I can accept these results, I have to attempt to corroborate

them by an independent mathematical analysis. This analytical problem would not have arisen had it not been for the electronic computer. The electronic computer is therefore, in this case, not a detractor of mathematical analysis, but serves rather like an experimental laboratory producing topics for the analysts to explain.[11]

It is likely that raising this argument was a preemptive step of Pekeris against prospective voices pushing for a change in orientation at DAM. The electronic computer—he meant to emphasize—rather than a limiting factor should be seen as the right means to broaden the domain of activities of the department. He sided with the leading advocates of the computer at the time, such as von Neumann and Ulam, who argued for the viability of "using the computers heuristically in the hope of obtaining insights in some questions of pure mathematics."[12]

Pekeris, however, ultimately failed in achieving that kind of diversification of interests. Indeed, in the last months of 1964, the Board of Governors of WIS put in place an external committee of experts and trusted them with the task of reviewing the various departments of the institute, while focusing on specific issues that had given rise to debate and criticism throughout the years. In particular, a subcommittee was appointed to focus on the activities of DAM and specifically on what some critics saw as a relatively narrow disciplinary scope. The two members of this subcommittee were Abraham Haskel Taub (1911–1999), a distinguished mathematical physicist from the University of Illinois, and Frank Press, already mentioned above, a distinguished geophysicist from CalTech and director of the SeismoLab since 1957. Their task was to review the role of mathematics at WIS at large, the future of the computer program, and the nature and extent of the efforts pursued in geophysics in the various departments of WIS. In early 1967, the committee published its report, suggesting changes to be considered for DAM. In view of the respective fields of interest and expertise of Taub and Press, this report reflects a surprisingly critical attitude, claiming that the research areas pursued at DAM were not sufficiently diverse and broad *within* the field of applied mathematics, while, at the same time, indicating that various areas of pure mathematics should also be practiced so that a much broader portion of current mathematical activities be covered. Moreover, they also found the computer program itself unsatisfactory and problematic. The report read as follows:

> Its programs essentially reflect the tastes and interests of its extremely able and distinguished director. With a highly competent and strong-willed scientist as leader, the major effort of the department at the moment is in theoretical numerical geophysics and in the design and construction of electronic computers. There is also important work on the wave functions of the helium atom.

---

[11] Pekeris, "Survey of research done by department of applied mathematics at Weizmann Institute of Science during the period of 1949–1956" (CPA).

[12] Stanislaw Ulam, "The computer in mathematics," Records of the Office of the Director/Faculty Files/Box 34; John von Neumann, John, "Conference: The Computer and the Development of Science and Learning," Folder 2. The Shelby White and Leon Levy Archives Center, Institute for Advanced Study, Princeton, NJ, USA (May 1972). Retrieved from https://albert.ias.edu/handle/20.500.12111/2687 (Jan 11, 2022).

... we are convinced that to make a continuing impact in this field the Institute requires greater diversification. It has to embark upon an experimental program and it should enter fields other than seismology and geophysics. It is difficult to justify the present deployment of such large human and financial resources to so restrictive a class of problems.

... the present applied mathematics of the institute ... is very good but highly specialized ... From the viewpoint of Israel as a whole, a good diversified applied math group would be a desirable addition. It would complement the strong pure math groups at the [Hebrew] University and the others now growing up at the Technion and other schools in the country. An applied math group would be less competitive with the other schools.

From the point of view of the Institute itself, even a broadly based applied math group would probably not be adequate ... it would be desirable to set up a single mathematics group containing a broad spectrum of pure and applied mathematicians ...

There are some members of the Survey Committee that have an even stronger feeling about this point. There is no pure mathematics department at the Institute and it means there is not a group at the Weizmann Institute responsible for imparting to graduate students basic knowledge in the variety of fields that they need to know in order to become up-to-date workers in some areas of applied mathematics, computer science and theoretical physics.[13]

As for the computer projects at WIS, although the reviewers acknowledged the importance of the work achieved thus far with WEIZAC and GOLEM, they were less enthusiastic about the future. They listed a few problematic points that required further discussion, while questioning the effectiveness of the computer program in terms of money, effort, and space. They listed five main problems:

1. The request for very high precision, which is the only justification for building a computer instead of buying one, will probably decrease in most of the departments of the institute.
2. No work on "online"[14] system is going on or planned.
3. There is almost no cooperation with the industry and educational institutes.
4. The resources that are required for the crucial advanced fabrication techniques are underestimated.
5. The software development efforts do not match those of the hardware development. For example, the GOLEM A FORTRAN compiler was completed two years after the completion of the machine.[15]

And, indeed, in 1969 a few changes were made. Shimon Even (1935–2004) came to WIS from Harvard University, were he had completed a PhD and was the visiting professor during the preceding two years. He was appointed to a chair in computer science, and he prepared the first specifically conceived graduate course taught in Rehovot. Also, in 1970, the Department of Pure Mathematics was established with a high-level graduate program in mathematics and new research areas. In addition,

---

[13] Report of the Weizmann Institute Survey Committee, Feb. 3, 1967 (WIA).

[14] The members of the committee did not specify what they meant when speaking about "online work." We believe that they had in mind something along the lines of the interactive work with the computer of Douglas C. Engelbart and his team, as eventually presented in Dec. 9, 1968, in the famous "Mother of All Demos." See https://www.dougengelbart.org/content/view/374/#8.

[15] The Scientific Activity Report, 1969 (WIA), indicates that the completion of the compiler will be delayed in about one year.

DAM appointed Isaac Horowitz (1920–2005) to a chair of system engineering, aimed at supporting the technological industries in Israel by providing *engineering-science training for engineers*.[16] Horowitz held a D.E.E. from the Polytechnic Institute of Brooklyn and had both industrial and academic experience in both Israel and the USA.

In fact, the archival record shows that debates about the possibility of creating a department of pure mathematics at WIS had started much earlier. Thus, for instance, in a visibly enraged letter that Pekeris sent in 1965 to the scientific director of WIS, the chemical physicist Shneior Lifson (1914–2001).[17] The letter was written after many heated discussions and efforts on the side of Pekeris to clarify his strong opposition to such a move. DAM had created—Pekeris argued—its own style of mathematics, a "strictly applied one" that would be "wiped out in the future." Pekeris emphasized that by having left pure mathematics outside the institution, WIS had been spared the "inevitable intrigues between pure and applied mathematics which have rocked even the oldest universities," including the likes of UCLA, CalTech, Cornell, MIT, Harvard, and Columbia. Pekeris would consider the introduction of pure math to be "a censure of our accomplishments" over the past eighteen years, and he expected that such a step would "do immediate harm to the research work" of DAM.[18]

But in 1969, Pekeris seems to have realized that establishing a group or a separate department of pure mathematics would become a reality, despite his own opinions. At the time, he may have still expected that it would be devoted mainly to teaching, whereas research would continue to focus on the applied agenda that he had been promoting all along. Pekeris wrote that:

> ... with the growth of Feinberg Graduate School [at WIS] there arose the need to provide instruction in the field of pure mathematics, in addition to applied mathematics, in which we have been active for the past 25 years. We were instrumental in organizing a new department of Pure Mathematics. This new Department will be headed by Professor S. Karlin of Stanford University. Professor S. Sternberg of Harvard University has also joined on part-time appointment. There exists, of course, a strong Pure Mathematics group at the Hebrew University in Jerusalem, but if we can maintain the standard of Professors Karlin and Sternberg, our efforts will be justified."[19]

During these years, great amounts of resources were spent on building the GOLEM B computer. However, it was also decided to purchase an IBM computer for the use of the entire scientific community of WIS, and though the GOLEM B project continued to its completion in 1974, this was another lost battle for Pekeris.

The department of pure mathematics became fully operative in 1971, as part of a faculty of Mathematics that also comprised the DAM. Samuel Karlin (1924–2007) was known for his broad range of interests, covering an astonishing variety of mathematical disciplines, including mathematical economics, game theory, evolutionary

---

[16] Scientific Activity Report, 1969 (WIA).

[17] Pekeris to the "Scientific Director of Weizmann Institute," Jan. 25, 1965 (CPA).

[18] Pekeris to the "Scientific Director of Weizmann Institute," Jan. 25, 1965 (CPA).

[19] Scientific Activity Report, 1969 (WIA).

theory, biomolecular sequence analysis, and matrix theory. He was hardly the mathematician typically identified with "pure" traditions of research. The many fields of research of Shlomo Sternberg (b. 1936) covered several classical areas of pure mathematics as well as mathematical physics.

The main declared goal of the two co-directors was to establish a high-level graduate program, with a special emphasis in areas such as algebra, real and complex analysis, functional analysis, probability, statistics, and mathematical biology. There was also a declared intention to develop, in addition to the classical fields of pure mathematics, new kinds of interdisciplinary projects, which fitted Karlin's own fields of interest well: theoretical population biology, quantitative aspects of evolutionary theory, quantitative and statistical genetics, demography, optimization methods in selection experiments, epidemiology, and ecology.[20]

Neither of these two prominent, well-established mathematicians ever left his position in the USA in order to settle in Rehovot, as Pekeris had done back in 1948. Still, with time, research in pure mathematics also reached world-class levels at WIS and the story of the department and its achievements certainly merits a detailed account of its own, but that would be well beyond the scope of the present one, and it remains to be done in a future opportunity.

References

Artstein, Z. 1995. Preface. In *Ocean Tides from Newton to Pekeris. The First Chaim Leib Pekeris Memorial Lecture*, ed. J.M. Lighthill, 5–9. Jerusalem: The Israel Academy of Sciences and Humanities.

Corry, L., and R. Leviathan. 2019. *WEIZAC: An Israeli Pioneering Adventure in Electronic Computing (1945–1963)*. Berlin: Springer.

Lighthill, M.J. 1995. *Ocean Tides from Newton to Pekeris. The First Chaim Leib Pekeris Memorial Lecture*. Jerusalem: Israel Academy of Sciences and Humanities.

Pekeris, C.L. 1972. Adventures in applied mathematics. *Quarterly of Applied Mathematics* 30 (1): 67–83.

Pekeris, C.L., and Y. Accad. 1969. Solution of Laplace's equations for the M 2 tide in the world oceans. *Philosophical Transactions of the Royal Society of London. Series A, Mathematical and Physical Sciences* 265 (1165): 413–436.

Pekeris, C.L., and K. Frankowski. 1993. Matching the Kerr solution on the surface of a rotating perfect fluid. *General Relativity and Gravitation* 25: 603–612.

Pnueli, A., and C.L. Pekeris. 1968a. Free tidal oscillations in rotating flat basins of the form of rectangles and of sectors of circles. *Philosophical Transactions of the Royal Society of London. Series A, Mathematical and Physical Sciences* 263: 49–171.

Pnueli, A., and C.L. Pekeris. 1968b. Tides in oceans in the form of a cross. *Proceedings of the Royal Society, Series A* 305: 219–233.

Shahar, A. 2002. *At the Frontline of Computing—Mamram: Legacy of the IDF Computers Center*. Tel Aviv: Maarachot [In Hebrew].

---

[20] Scientific Activity Report, 1971 (WIA).

Chapter 8
Concluding Remarks

Abstract Chaim L. Pekeris was a visionary scientist-entrepreneur and broad-minded, versatile applied mathematician. He convinced the leadership of the Weizmann Institute in Rehovot to embark on the unlikely project of building an electronic, high-speed computer in the newly created State of Israel. He led the project to completion and undertook the task of approaching complex scientific questions in various fields. He came up with ground-breaking results based on the use of WEIZAC and thus led the creation of the technical and human infrastructure on which the computer culture was to thrive in Israel in the forthcoming decades.

Keywords WEIZAC · Chaim L. Pekeris

This book has discussed the scientific contributions of Pekeris and his collaborators at WIS that were based on calculations performed with WEIZAC. We analyzed the historical background and the impact of the disciplines where these contributions took place, mainly integral equations, geophysics, spectral theory of many-electron atoms, and atomic spectroscopy. Pekeris, a visionary scientist-entrepreneur and broad-minded, versatile applied mathematician, was able to convince the leadership of the Weizmann Institute in Rehovot to embark on the unlikely project of building an electronic, high-speed computer in the newly created State of Israel. He led the project to a prompt and successful completion and then undertook the task of approaching complex scientific questions in various fields and coming up with groundbreaking results based on the use of WEIZAC. He almost singlehandedly led the creation of the technical and human infrastructure on which the computer culture was to thrive in Israel in the forthcoming decades.

The foregoing account, as already indicated, is a follow-up to our previous publication, *WEIZAC: An Israeli Pioneering Adventure in Electronic Computing (1945–1963)*. These two texts are intended as complementary to each other, and they are best read in conjunction. They help understand how WEIZAC came to be the starting point of the creation of an autonomous, multifaceted computing culture in Israel, relatively soon after the State of Israel was born. This culture would come to include the adoption of electronic computers in government institutions, and in particular—in a most significant way—by the military and by the intelligence communities, their increased use in scientific research in various disciplines, the rise of a well-developed

© The Author(s), under exclusive license to Springer Nature Switzerland AG 2023   119
L. Corry and R. Leviathan, *Chaim L. Pekeris and the Art of Applying Mathematics with WEIZAC, 1955–1963*, SpringerBriefs in History of Science and Technology,
https://doi.org/10.1007/978-3-031-27125-0_8

high-tech industry, and the consolidation of a world-class, autonomous new discipline of computer science in the country. Concerning the latter issue, we would at least like to point out in passing that, though the crucial contribution of WEIZAC to the development of science in Israel and in WIS is unquestionable, the development of computer science as an academic discipline in WIS, and beyond, was promoted or comparatively held back by Pekeris's attitudes and charismatic authority.

Much as the WEIZAC project itself was the starting point of the story and Pekeris was the main moving force that led to its materialization and successful use, our two publications signal only a starting point of a broader historical enquiry about the story of the Israeli computing culture, and it will be continued in a series of planned, future publications.

Index

© The Author(s), under exclusive license to Springer Nature Switzerland AG 2023   121
L. Corry and R. Leviathan, *Chaim L. Pekeris and the Art of Applying Mathematics with
WEIZAC, 1955–1963*, SpringerBriefs in History of Science and Technology,
https://doi.org/10.1007/978-3-031-27125-0

Printed in the United States
by Baker & Taylor Publisher Services